DEVICES FOR OPTOELECTRONICS

OPTICAL ENGINEERING

Series Editor

Brian J. Thompson

Distinguished University Professor
Professor of Optics
Provost Emeritus

University of Rochester
Rochester, New York

Additional Volumes in Preparation

DEVICES FOR OPTOELECTRONICS

WALLACE B. LEIGH

Alfred University
Alfred, New York

Marcel Dekker, Inc.   New York•Basel•Hong Kong

Library of Congress Cataloging-in-Publication Data

Leigh, Wallace B.
    Devices for optoelectronics / Wallace B. Leigh.
        p.   cm. — (Optical engineering ; v. 55)
    Includes index.
    ISBN 0-8247-9432-X (hardcover : alk. paper)
    1. Optoelectronic devices. 2. Optoelectronics. 3. Solid state physics.
    4. Optical communications—Equipment and supplies. I. Title. II. Series:
Optical engineering (Marcel Dekker, Inc.) ; v. 55.
    TA1750.L44 1996
    621.3815'2—dc20   96-17879
                                                                                  CIP

The publisher offers discounts on this book when ordered in bulk quantities. For more information, write to Special Sales/Professional Marketing at the address below.

This book is printed on acid-free paper.

Marcel Dekker, Inc.
270 Madison Avenue, New York, New York 10016

Current printing (last digit):
10 9 8 7 6 5 4 3 2 1

PRINTED IN THE UNITED STATES OF AMERICA

From the Series Editor

The language and techniques of optical science, engineering, and technology have changed significantly in recent years because systems have become hybrid. They are hybrid in the sense that systems are made up of passive and active optical subsystems, sources, beam manipulators, detector arrays, computer subsystems, optical and electronic image and signal processing subsystems, and so on. Optical devices are activated by electrical, acoustical, and magnetic signals to provide switches, modulators, scanners, isolaters, and connectors.

In this book, Wallace B. Leigh addresses the important subject of devices for optoelectronics. Optoelectronics may not enjoy a universally accepted definition. One source* defines optoelectronics as "pertaining to a device that responds to optical power, emits or modifies optical radiation, or utilizes optical radiation for its internal operation; any device that functions as an electrical-to-optical or optical-to-electrical transducer." Other writers use the term synonymously with electro-optics or photonics, but let us not get into that semantic discussion. Suffice it to say that optoelectronics involves the interface between light and electronics and is a valuable subfield of optical science and technology.

Dr. Leigh covers a range of appropriate topics, starting with some vital fundamentals on the properties of materials that are essential for these

---

The Photonics Dictionary, Lauren Publishing, Pittsfield, MA, 1995.

devices. Since laser light sources are important to these systems, semiconductor lasers are covered. Light-emitting diodes are also valuable light sources and are covered in a separate chapter. These two types of sources have the advantage of being easily coupled into components and subsystems. Overview of detectors and acousto-optic, electro-optic, and liquid crystal devices complete the volume to produce a very fine introduction to this subject.

Brian J. Thompson

Preface

The transmission and processing of information at optical frequencies is fast becoming a dominating technology in communications and information processing. The potential advantages behind high-bandwidth, low-noise optical systems encouraged the United States, Japan, and Europe to make large-scale optoelectronic processing a national goal more than 15 years ago, and a large optoelectronic research program exists in almost every major country of the world. Optical communication systems will rely on a wide variety of optoelectronic devices for information processing, switching, modulation, optical memories, etc. Many of these devices will be integrated with optical waveguides into optical integrated circuits (OICs), which perform logic functions using optical signals with or instead of electrical signals. A crucial factor in the success of optical communication systems is in the education of scientists and engineers working in the field of optoelectronic devices. This book is for people designing optical communication systems, doing research in the area, or teaching a first-year graduate course.

At present, for technical work concerning the physics of optoelectronic devices, a student of optoelectronics must turn to (1) a general-purpose book on the physics of semiconductor devices, (2) a book on general optoelectronics or fiber optics, or (3) conference proceedings. A general book on the physics of semiconductor devices focuses primarily on the physics of active devices for electronics (silicon diodes and transistors), with extra chapter or a section at the end of a chapter added for optoelectronic devices. General books on optoelectronics are more concerned with the systems aspect of optical communication systems, and less involved with the technical aspects of the physics of optoelectronic devices. Conference proceedings include mainly those published by the Society of Photo-Optical Instrumentation Engineers (SPIE) or the Optical Society of America. While conference proceedings contain valuable information on detailed subjects, no generic information of optoelectronic device physics can be obtained from them.

This book focuses directly on the physics of optoelectronic devices. Furthermore, the focus is on devices that may eventually see use in either hybrid or monolithic optical integrated circuits. Such optoelectronic devices can be broken down into three basic categories: optical transmitters, optical receivers, and optical switching devices.

Transmitters include low-power devices, including light-emitting diodes (LEDs), laser diodes (also known as semiconductor lasers), and display devices. LEDs and laser diodes have undergone enormous development in the last few years. As an example, in consumer electronics, compact disk players are now more reliable than ever thanks to progress in laser diode technology. Chapters 2 and 3 are dedicated entirely to these two devices, which have become so important.

Liquid crystal display (LCD) devices (Chapter 9) are also important technologies. Most of the progress made in this area has been outside the consumer arena; however, an example of LCD technology is currently underway in flat-screen television, yet another example of optoelectronics technology that has touched the consumer products industry.

Receivers convert light to electricity and include pin diodes and avalanche photodiodes. Chapter 4 on photodetectors includes information on charge coupled devices and array detectors.

Few technical monographs cover both active semiconductor devices (transmitters and receivers) and optical switches. Switching devices—optoelectronic devices used in thin-film waveguides—include electrooptic and self-electro-optic effect devices, among others. Chapter 5 bridges the gap in film switches by including hybrid (lithium niobate) as well as active (compound semiconductor) devices. There is also a chapter on nonlinear devices for OIC applications.

This book is concerned only with optoelectronic devices used in communication and information processing, i.e., with low-power devices that may eventually be integrated into optical integrated circuits. Devices used for control of high-power optics such as laser modulators, isolators, and Q-switches (sometimes considered devices for optical engineering) are not included. The physics of some of these devices may be similar to devices for optical engineering, but it is time to differentiate between them. Devices for photovoltaics or solar cells are also not included, as there are many technical monographs in this area as well.

Devices for Optoelectronics describes several new devices that will be important to the field of optoelectronics in the next few years. In particular, the laser diode, LCD, and thin-film devices will continue to progress, and more optical and optoelectronic integrated circuits will appear in research labs.

Wallace B. Leigh

Contents

DEVICES FOR OPTOELECTRONICS

1

Properties of Optoelectronic Materials

1 INTRODUCTION

A reliable optoelectronic device must meet specified criteria for optimal performance. Among other things, the device needs to have the proper spectral response for the optical wavelength of interest, the proper speed of response, and be fabricated at a reasonable cost. These are all basic material-dependent concerns, and quite often the materials from which the device is fabricated determine whether the device meets these requirements.

An important aspect is the ability to process the material. Many proposed electronic and optical devices have failed commercially because of difficulties in finding a suitable material which can be easily processed with an acceptable device yield. Optoelectronic device processing is a vast and important subject in itself and is beyond the scope of this text.

In this chapter we will review properties of materials used for the fabrication of optoelectronic devices. We are concerned with how electric fields and currents interact within these materials and how light interacts with the solid. We will observe how light interacts with current carriers such as free electrons or free holes. We will study the energy bands of semiconductors under ideal and nonideal conditions and observe how optical properties of materials change under external perturbations such as electric fields or pressure or temperature gradients.

2 ENERGY BAND STRUCTURE

To obtain a basic understanding of the electronic nature of semiconductors, it is appropriate to begin by examining the de Broglie relation. This relation is based on the realization that particles such as electrons have a wavelength λ which can be related to their momentum p by the relationship

$$\mathbf{p} = \hbar\mathbf{k}$$

where $\hbar$ is the reduced Plank's constant $(h/2\pi)$ and $|\mathbf{k}| = k$ is the wavenumber $(2\pi/\lambda)$. An electron in a solid will exhibit particle-like properties; it will also behave as a wave capable of diffracting off of a grating of atomic dimensions. The set of points which represents the positions of the atoms in the crystal, called a *Bravais lattice*, is just such a grating. Thus, physical parameters, such as the mass of the electron and its contribution to the conductivity, must be considered within the framework of both the wave and particle aspects of the electron.

It is also important to understand that the electronic nature of the solid is not complete unless one examines the energy structure of the solid as a whole, and not just at a few atoms at a time. The valence electrons interact over several atomic potentials in the crystal. Since the entire crystal is involved, we take advantage of the symmetry of the lattice when describing physical properties of that crystal. Electronic properties, for example, are always described along lines of highest symmetry in the lattice. For this reason it is helpful to understand how symmetry is contained within a lattice.

The Bravais lattice of silicon is a diamond lattice, which has a symmetry similar to the face-centered cubic (fcc) structure as demonstrated in Fig. 1. The diamond lattice can be constructed from two superimposed fcc cubes, the corner of one cube located along the body diagonal of the other cube but displaced by a distance $(\frac{1}{4}, \frac{1}{4}, \frac{1}{4})$ along that diagonal. Compound semiconductors used in optoelectronic devices, such as GaAs and InP, have the zincblende structure. This structure is similar to the diamond and is the same lattice except that one of the fcc cubes of the lattice is occupied by the cation of the compound, and the other by the anion. (For example, all gallium on one cube and all arsenic on the other for GaAs.)

The smallest translatable cell in the diamond lattice is one that contains only one atom and is called the primitive cell. The primitive cell for diamond is a parallelepiped and is shown in Fig. 1. The three principal axis of the parallelepiped represent the three primitive, or shortest, translation vectors along which the cell can be translated to form the entire three-

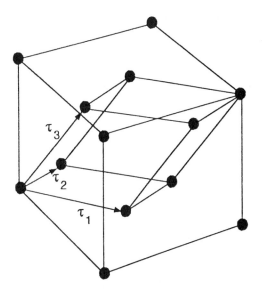

Figure 1 Diamond lattice, the atomic structure of elemental semiconductors Si and Ge. The lattice can be represented by two interlocking face-centered cubic structures. Primitive cell for a diamond lattice is demonstrated by the parallelepiped formed by vectors τ_1, τ_2, and τ_3.

dimensional lattice. Translation groups are vectors which are represented by the sum of integral multiples of the primitive translation vectors:

$$\mathbf{T} = n_1\boldsymbol{\tau}_1 + n_2\boldsymbol{\tau}_2 + n_3\boldsymbol{\tau}_3 \tag{1}$$

where the τ_i are the primitive translation vectors and the n_i are integers. Any set of n_i leaves the crystal invariant. The set of points connected by the primitive translation vectors is the Bravais lattice. All allowed rotations, reflections, translations within the Bravais lattice, which keep the lattice invariant, are called space groups. Space groups with the translational portion omitted are called point groups. A group is a collection of symmetry operations which takes not just the lattice but also the Hamiltonian representing the atomic potentials into itself.

We learn of the structure of the electronic states of a solid by studying how they transform under a symmetry operation, T. Translation of a one-dimensional function $f(x)$ by a vector T displaces the function by an amount $x + a$:

$$Tf(x) = f(x + a) = \lambda_k f(x) \tag{2}$$

Here λ_k is the matrix that represents the symmetry operator T. If we start, in one dimension, with a ring of atoms, the periodic boundary conditions are cyclic, and we can write

$$\lambda_k(T) = \exp\left(\frac{2\pi i n}{N}\right) = e^{ika} \tag{3}$$

where we have N atoms spaced a distance a in the chain, and n is an integer chosen to run from $-N/2$ to $N/2$. The wavenumber k is defined as

$$k = \frac{2\pi n}{Na} \tag{4}$$

The boundary conditions restrict k to values of $\pm \pi/a$. There are N translations in this group and N values of n which give distinct representations.

If the wavefunction of an electron is ψ, we wish to be able to make a transformation on ψ such that

$$T\psi_k(x) = e^{ika}\psi_k(x) = \psi(x + a) \tag{5}$$

We now define a new function $u_k(x)$ such that

$$u_k(x) = u_k(x + a) \tag{6}$$

and

$$\psi_k(x) = u_k(x)e^{ikx} \tag{7}$$

which yields

$$\psi_k(x + a) = u_k(x + a)e^{ik(x + a)} = u_k(x)e^{ika}e^{ikx} = e^{ika}\psi_k(x)$$

and note that $Tu_k = u_k$, or the function is invariant under the symmetry operation. The wavefunctions are now given by Eq. 7, and the functions u_k are known as Bloch functions, and have the full translational symmetry of the lattice.

If u_k is a constant, the wave function for each k is simply a plane wave which satisfies the periodic boundary conditions. One approach to understanding electron wave functions in a solid is to treat them as plane waves which are modified through the functions u_k. The presence of the atoms enters through the u_k function. Figure 2 illustrates the basic parts of wave functions as represented by Eq. 7.

When $n = N/2$, $k = \pi/a$ and k is at its maximum allowed value, which is at the edge of the first Brillouin zone. The first Brillouin zone is the volume of k space containing all values of k between $\pm \pi/a$. Any value

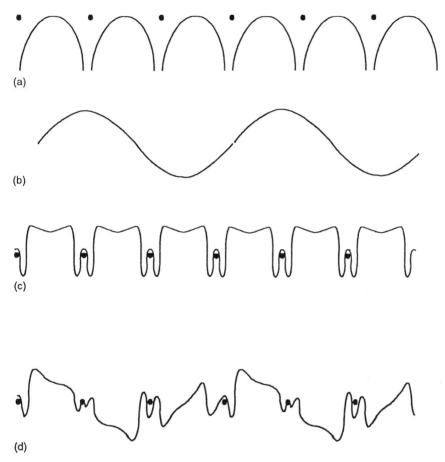

Figure 2 Construction of Bloch waves from individual components, as represented along a row of atoms. (a) The Coulombic potential of the row. A plane
wave (b) and the Bloch function (c) are combined to form a wavefunction (d) for
the row.

of $k' > \pi/a$ moves the electron into the next Brillouin zone, and eigenstates
of the Schrödinger equation for this value of k can be represented by a
reduced form through the translation

$$k = k' - \frac{\pi}{a}$$

and thus the eigenstates are transformed back to the first Brillouin zone. The eigenstates are so close to one another in energy ($\sim 10^{-18}$ eV) that these states are usually treated as a continuum known as energy bands.

It is helpful to observe a band structure of a two-dimensional lattice, since we can use the third dimension to plot energy. Hypothetical energy band representations for a square lattice of spacing a are shown in Fig. 3. The Brillouin zone is also a square, shown in Fig. 3a. The primitive lattice vectors have magnitude $2\pi/a$. The bands are also shown in two-dimensional E versus k plot in Fig. 3b, along values of k of highest symmetry, from W to Γ to X and then back to W. This corresponds to lines running from a corner of the square to the center, to the center of an edge and back to the corner.

Plotting energy bands in this manner, along lines of highest symmetry, will also allow us to detail the bands of a three-dimensional lattice. For the hypothetical energy bands represented in Fig. 3b, the minimum in the valence band is at the same point as the maximum in the conduction band, and the semiconductor is said to possess a *direct gap*. While a direct band gap is represented in the two-dimensional-lattice band structure depicted in Fig. 3, in a real system indirect band gaps are also possible.

The diamond lattice has the same Brillouin zone and symmetry lines as the face-centered cubic structure. The Brillouin zone for silicon is illustrated in Fig. 4. Principal directions are shown in the zone where the atomic planes have four-, three, and twofold symmetry, namely the [100], [111], and [110] directions, respectively. These will be the principal directions by which the energy bands will be plotted.

The features of the energy band edges surrounding the forbidden energy gap are similar for all diamond and zincblende semiconductors. The bands are plotted along high symmetry lines, that is, [100], [110], and [111] directions. The [100] direction is represented by the Δ axis, the [111] by the Λ axis, and the 110 by the Σ axis. Note that K and U are equivalent points on the Brillouin zone. If one starts at Γ and travels along the Σ direction past K (or U), one will end at point X in a neighboring zone. Thus, the energy bands for diamond and zincblende lattices are drawn from L to Γ to X and then back to Γ. The path from X back to Γ is outside the first zone, but can be translated into the first zone along a translational vector. The energy gap is defined by the difference in energy between the lowest conduction band valley and the valence band maximum at Γ. For diamond and zincblende semiconductors there are conduction band minima somewhere along all three of these directions, but a valence band maximum only exists in the zone center at Γ. Energy bands plotted along these same directions for GaAs and InP are shown in Fig. 5.

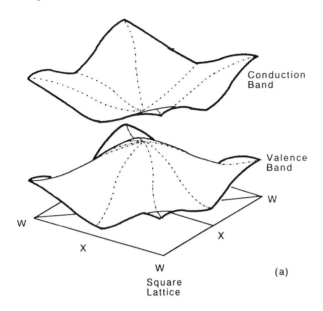

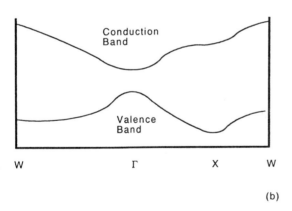

Figure 3 Energy bands of a 2D lattice (after Ref. 1). (a) Sketch of a hypothetical energy band edge as a function of k. The 3D plot shows how the energy, plotted along the verticle axis, varies across a square Brillouin zone. W, X, and Γ represent points of highest symmetry (corners, side, and center) of the square. (b) The conventional way of describing the same bands, plotted only along the directions of highest symmetry. The figure shows, for illustrative purposes, a direct band gap at Γ.

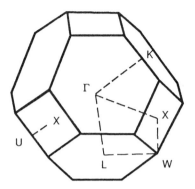

Figure 4 Brillouin zone of diamond and zincblende lattices [1]. The points Γ, K, and X are points of highest symmetry.

A conduction band minimum and valence band maximum that occur at the same k value produce a direct gap. In Ge the direct gap is only 0.16 eV larger than the indirect gap, and in Si it is ~1.2 eV larger, while the lowest energy gaps for GaAs and InP are direct. In fact, all III-V compounds except Al-containing compounds and GaP have conduction band minima at $k = 0$, and the direct gap is the smallest gap for these compounds. Direct gaps are also found for the II-VI compounds ZnSe, ZnTe, ZnS, CdSe, CdS, and CdTe. Direct gaps are of importance in order to understand such phenomena as absorption and radiative recombination in Chapter 2.

In the valence band structure of III-V compounds, the twofold spin degeneracy of the V1 band that exists in Si and Ge is lifted due to the nearest-neighbor atoms being of a different element. There is also a similar small lifting of the V2 band. This is illustrated in Fig. 5 as a "split" between the points Γ_8 and Γ_7. Since electrons (and holes) have an effective mass dependent on band curvature, as discussed in Section 4, holes in these bands are of "light" or "heavy" mass, depending on which band the hole resides in. Some devices exploit this difference in the hole effective mass.

In a three-dimensional energy band representation, the energy bands form closed shells which represent constant-energy surfaces. If a line of constant energy is placed slightly above the lowest [100] conduction band valley of GaAs, for example between 1.4 and 2.0 eV in Fig. 5, surfaces can be traced by plotting points of constant energy in the three dimensions. The surfaces formed for silicon or GaAs along their lowest minima are

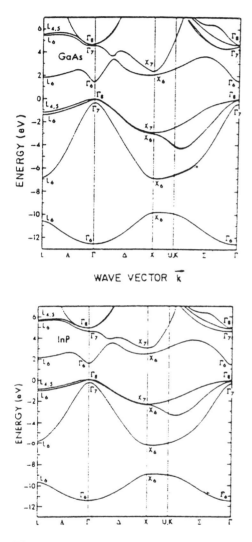

Figure 5 Energy bands of GaAs and InP [2].

six ellipsoids that exist along each of the [100] directions. These surfaces are shown for silicon in Fig. 6. With each increment in k there is a quadratic increase in energy along each direction, but with a differing constant of proportionality in each of the three, which traces an ellipsoid. For any constant energy close to a conduction band minimum or valence band maximum, ellipsoids will exist in the constant-energy representation.

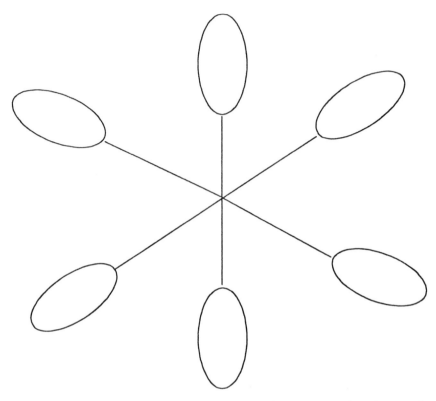

Figure 6   The six constant energy ellipsoids of an energy value close to the edge of a near-parabolic band. Such a band is found in the conduction energies of silicon.

These areas of the bands represent where the structure most closely resembles the free-electron theory.

There are six ellipsoids along [100] directions in silicon, since this minimum occurs within the first Brillouin zone (85% along the way from Γ to X). If the minimum occurs right at the Brillouin zone, then there are only half-ellipsoids in the constant-energy-surface representation. This is the case for the [111] minima of germanium. The conduction band minima falls right at the Brillouin zone boundary, and [111] Ge ellipsoids are cut along the short direction in the constant-energy representation. Thus there are only four [111] ellipsoids for Ge instead of the eight it would have if the minimum of the conduction band was completely within the Brillouin zone.

Depending on the interaction of nearest- and next-nearest neighbor atoms, the minimum in the conduction band may not occur at the Γ center, as it indeed does not for silicon. However, in the semiconducting alloy Ge_xSi_{1-x}, the nature of the lowest conduction band valley changes with composition. The [111] and [100] (both indirect) conduction band valleys of germanium both move to higher energies with increasing silicon content, but the [111] valleys move faster. The movement is not linear, and at one point, at approximately 15 mole% silicon, the two energy gaps are equal [3].

3 WAVE PACKETS AND ELECTRIC FIELDS IN SEMICONDUCTORS AND THE MOTION OF ELECTRONS

To further illustrate the use of Bloch functions in wave dynamics, it is helpful to look at the electron as a wave packet. This helps describe the motion of an electron in the bands. The wave function can be expanded in a Fourier series around a central wavenumber k_0. The electron is localized at k_0 in one particular band. Summing the wavefunction over a series of wavevectors $\mathbf{k}$ parallel to $\mathbf{k_0}$ gives

$$\psi = \sum_k u_k \exp(i\mathbf{k}\cdot\mathbf{r})\exp[-\alpha(\mathbf{k}-\mathbf{k_0})^2]\exp\left[-i\left(\frac{E(\mathbf{k})}{\hbar}\right)t\right] \tag{8}$$

$$= \exp[i\mathbf{k_0}\cdot\mathbf{r}]\cdot\sum_k u_k \exp[i(\mathbf{k}-\mathbf{k_0})\cdot\mathbf{r}]\exp[-\alpha(\mathbf{k}-\mathbf{k_0})^2]\exp\left[-i\left(\frac{E(\mathbf{k})}{\hbar}\right)t\right] \tag{9}$$

This is a Gaussian packet centered at $\mathbf{r} = 0$ and modulated by $u_k e^{i k \cdot r}$. The Gaussian function is chosen in order to have a sharp maximum at $k = k_0$ and to fall off rapidly away from k_0.

Because of this property of the Gaussian, we can expand $E(k)$ about k in a Taylor series and retain the first two terms:

$$E(\mathbf{k}) = E(\mathbf{k_0}) + (\mathbf{k_0} - \mathbf{k})\cdot\nabla_k E(\mathbf{k}) \tag{10}$$

Using this expansion, $\psi(\mathbf{r},t)$ becomes

$$\psi(\mathbf{r},t) = \psi_0 \sum u_k \exp\left[i(\mathbf{k}-\mathbf{k_0})\cdot\mathbf{r} - i\left(\frac{1}{\hbar}\right)\nabla_k E(\mathbf{k})t\right]\exp[-\alpha(\mathbf{k}-\mathbf{k_0})^2] \tag{11}$$

where

$$\psi_0 = \exp\left[i(\mathbf{k}_0 - \mathbf{r}) - i\left(\frac{E(\mathbf{k}_0)}{\hbar}\right)t\right] \tag{12}$$

which is the same wave packet displaced by an amount $(1/\hbar)\nabla_k E(\mathbf{k})t$, or the packet moves with group velocity v_g:

$$v_g = \left(\frac{1}{\hbar}\right)\nabla_k E(\mathbf{k}) \tag{13}$$

This concept of an electron traveling as a packet with a group velocity is entirely analogous to that used in optics to define electromagnetic waves as photons.

The effect of applying an electric field in one direction on the electron in k-space can be determined from the group velocity of the electron wave packet and the change of energy with time:

$$\frac{dE}{dt} = \frac{dk}{dt}\frac{dE}{dk} = F \cdot v_g = \left(\frac{1}{\hbar}\right) F \cdot \frac{dE}{dk} \tag{14}$$

Along one direction

$$F = \hbar\frac{dk}{dt} = -e\mathscr{E} \tag{15}$$

where $\mathscr{E}$ is the electric field strength. The effect of the electric field is a linear motion in k-space, as shown in Fig. 7, with a velocity dk/dt which is constant and equal to $-(e/\hbar)\mathscr{E}$. As k varies in the zone, the energy of the electron varies with the curvature of the bands. If the electric field is applied along a line of symmetry, the wavenumber will progress through the zone center, the energy varying along a band, until the edge of the Brillouin zone is reached. At this point, due to the translational symmetry of k-space, the wavenumber disappears and reappears at the opposite Brillouin zone boundary. It repeats its first journey, returning through $k = 0$. This is the equivalent of having the electron, as a de Broglie wave, diffract off the array represented by the lattice. The electron will remain in the same band, unless scattered into a higher energy band by some mechanism, such as an optical absorption process. In Fig. 7a, the velocities in real space that are given by $(1/\hbar)(dE/dk_x + dE/dk_y)$ are shown as vectors on the k-space trajectory. The shape of the corresponding path in real space is shown in Fig. 7b. The path taken (with no scattering) assumes a real-space trajectory based on the vectors perpendicular to the k-space trajectories.

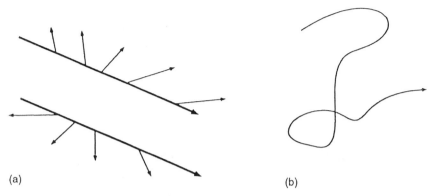

Figure 7 (a) Straight-line motion of carriers under an electric field in k-space. The vectors represent the velocity of the carriers in real space. (b) Trajectories of the electron in real space.

4 CARRIER LIFETIME AND EFFECTIVE MASS

In a real crystal the electron is scattered before it completes a full cycle through the Brillouin zone. Scattering mechanisms include scattering from lattice vibrations (phonons), impurities, or defects; that is, any deviation from the perfect symmetry of the lattice. The electron can also be scattered by the edge of the Brillouin zone boundary. Scattering in electron dynamics is included in a quantity called the carrier lifetime. There are several ways in which lifetimes can be defined, including a semiclassical approach:

$$F = \frac{m^* v_T}{\tau} \tag{16}$$

The quantity τ as defined here represents the average scattering time between collisions, v_T is the thermal velocity of carriers, and m^* is the effective mass.

The effective mass of a carrier can be determined by assuming that the band close to a valley in a conduction band, or that close to a hill in the valence band, is near parabolic. If we regard the E-versus-k curves in a solid as parabolic, as it is for the free electron, we must define an effective mass for differing curvature of the bands. We write, for the portion of the bands close to a band edge,

$$E(k) = \frac{\hbar^2 k^2}{2m^*} \rightarrow m^* = \frac{\hbar^2}{d^2 E/dk^2} \tag{17}$$

Defining the bands close to the band edge is appropriate, as the free carriers reside primarily within this region. For a free electron, the E-versus-k curves have the same curvature for all values of k since the mass m_0 is a constant. In a solid, however, the effective mass will vary along different directions of the Brillouin zone as the curvature of the band varies.

The six minima of the constant-energy surfaces in Fig. 6 each have a different curvature for the principal directions in the ellipsoid and, thus, different effective masses for these directions. For silicon, a longitudinal mass $m_L = 0.91m_0$ and transverse mass $m_T = 0.19m_0$ characterize the principal ellipsoid directions. For carrier statistics, the effective mass is defined by a single spherical surface with a density-of-states effective mass defined by

$$m_e = 6^{2/3}(m_L m_T^2)^{1/3} = 1.08m_0 \tag{18}$$

We still have, for group velocity,

$$v_g = \nabla_k \omega = \left(\frac{1}{\hbar}\right) \nabla_k E \tag{19}$$

The sign of the velocity is different for different $\nabla_k E$ of the $E(k)$ curves. We note, from where the curvature is negative, that m^* has regions in which it is negative. For $m^* < 0$, the acceleration due to the applied force is in a direction opposite to the applied force. This corresponds to the action of hole movement, as holes in the valence band move opposite to the direction of electrons. When speaking of holes, it is important to treat them as an entity distinct from electrons. Thus k for a hole state is the negative of the k of the missing electron state, and the energy of the hole is the negative of the energy of the missing electron. Under an electric field applied in the $+x$ direction, electrons are accelerated in the $-k_x$ direction and the current is in the $+x$ direction. If there is a hole in this band, under a field in the $+x$ direction, it moves in the $+k_x$ direction and the current is still in the $+x$ direction. The hole moves with a finite velocity as given by Eq. 19. The hole acquires a velocity in the direction of the applied field, consistent with the concept that it is a particle with positive charge.

5 DENSITY OF STATES

The values of k taken together represent a lattice of equally spaced points in k-space. The constant-energy surfaces in k space are, to a first approximation, spherical. The volume of a spherical shell in k-space having energy between E and $E + dE$ is $4\pi k^2 \, dk$. A single state occupies a cube in k-

space which has equal sides $2\pi/L$ and volume $8\pi^3/V$, where V is the volume of the sample having sides of length L. The number of these k-space cubes (or states) within the thin spherical volume of k-space per volume of crystal is $g(k)\ dk = k^2/\pi^2\ dk$. This number was multiplied by 2 to allow for two electrons (one of each spin) per cube. Using $E = \hbar^2 k^2/2m^*$ as the dispersion relation near the conduction band minimum, we obtain $dE/dk = \hbar^2 k/m^*$. We start with

$$g(k)dk = g(k)\frac{dk}{dE}\,dE \tag{20}$$

If $g(E)$ is the number of states per volume (density of states) between E and $E + dE$, we obtain by substitution that

$$g(E)\ dE = 4\pi \left(\frac{2m^*}{h^2}\right)^{3/2} E^{1/2}\ dE \tag{21}$$

Since the valleys are not exactly spherical, but more elliptical as shown in Fig. 6, m^* is not isotropic and we must account for this in the density of states effective mass by writing

$$m^* = (m_l^* m_{t1}^* m_{t2}^*)^{1/3} \tag{22}$$

where m_l^*, m_{t1}, and $*m_{t2}^*$ are the longitudinal and transverse effective masses along the ellipsoids of Fig. 6. Assumption of parabolic bands is only a first approximation to obtaining Eq. 22, and suitable corrections can always be made to account for any nonparabolic features of the bands.

A set of states is contributed from each valley in the band. Each energy level then may actually have contributions from several valleys. Summing the density of states of all states gives

$$N = \frac{1}{3\pi^2 h^3} \sum g_j (2m^*)^{3/2} (E - E_j)^{3/2} \tag{23}$$

where the g_j are the number of valleys of type j. Attempting to correct for the contribution from all valleys, as well as correcting for the nonparabolic nature of the bands, is a somewhat arduous task.

If we plot density of states versus energy given by Eq. 21, the results are shown in Fig. 8a. It is important to note that the density of states goes to zero as one moves toward the edges of the bands. This is important when discussing luminescence or absorption mechanisms. If electrons make the transition between the very lowest level of the conduction band and the topmost level of the valence band, a very high flux of these transitions may be limited by the somewhat limited density of states found there.

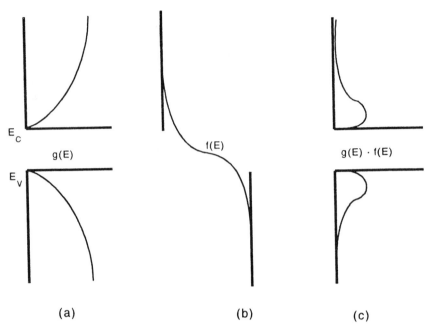

Figure 8   (a) Density of states ($g(E)$) vs. energy, (b) Fermi function as a function of energy, and (c) the resulting distribution of carriers in the conduction and valence band formed from the product of (a) and (b).

6   CARRIER CONCENTRATION

The above expressions for the density of states tells us only how closely spaced together the energy levels are in the bands. They do not inform us if these levels are occupied. The probability of occupancy of a set of levels is dependent on statistics and varies with temperature of the solid. Occupational probabilities are given by the Fermi function:

$$f(E) = \frac{1}{\exp\{(E - E_F)/kT\}} \tag{24}$$

The probability that the level is not occupied by an electron, and hence occupied by a hole, is given by

$$1 - f(E) = \frac{1}{\exp\{(E_F - E)/kT\}} \tag{25}$$

The Fermi energy, E_F, is measured from the bottom of the bands upward for Eqs. 24 and 25. The parameter E_F is defined as the energy where electron and hole occupancies are equal; that is, $f(E_F) = 0.5$. Because of the exponential nature of the Fermi function, there is only a very small (about kT) distance below the Fermi level where the electron occupancy probability rises to unity, and an equally small distance above E_F where it falls off to zero.

The total number of electrons is determined by the number of available levels at the energy of the interaction—that is, the density of states—multiplied by the probability that the level is occupied, and summed over all levels. The electron concentration n in the conduction band can be found from integrating over the product of $g(E)$ and $f(E)$ over the range of bandgap energies, from the bottom of the band at energy E_c upward:

$$n = \int_{E_C}^{\infty} f(E)g(E)\, dE \tag{26}$$

The function $g(E)$ increases as $E^{1/2}$ while the function $f(E)$ sharply decreases with increasing energy. The product $f(E)g(E)$ is a distribution of electrons with energy into a band that peaks a very short distance ($\sim kT/2$) from the band edge. Figure 8c shows how this distribution is shaped. The result is that most free carriers at normal temperatures reside very closely to a band edge.

Solutions to the integral in Eq. 26 are most easily found if we assume that the material is only weakly doped; that is the Fermi energy is at least a few kT from either band edge, and thus we can approximate the Fermi function as

$$f(E) = e^{-(E_c - E_F)/kT} \tag{27}$$

and the concentration of electrons can be determined:

$$n = N_c e^{-(E_c - E_F)/kT} \tag{28}$$

where N_c, defined as

$$N_c = 2 \left(\frac{2\pi m_n^* kT}{h^2} \right)^{3/2} \tag{29}$$

is known as the effective density of states for the conduction band, and m_n^* is the effective mass of the electrons in that band. N_c is equal to the electron concentration obtained if the Fermi energy resides at the conduction band edge, i.e., when $E_F = E_c$. For holes in the valence band, we can write a similar statement and show that

$$p(E_F) = N_v e^{-(E_F - E_v)/kT} \tag{30}$$

$$N_v = 2 \left(\frac{2\pi m_p^* kT}{h^2} \right)^{3/2} \tag{31}$$

If we take the product of Eqs. 28 and 30 we have

$$p \cdot n = N_c N_v e^{-(E_c - E_v)/kT} = N_c N_v e^{-E_g/kT} \tag{32}$$

where $E_g = E_c - E_v$ is the band-gap energy. We can define an intrinsic energy E_i such that $E_F = E_i$ when $n = p = n_i$. We call n_i the intrinsic pair density or the intrinsic concentration. Setting n of Eq. 28 equal to p of Eq. 30 gives

$$E_c - E_i = \frac{E_g}{2} + \frac{3kT}{4} \ln\left[\frac{m_p^*}{m_n^*} \right] \tag{33}$$

and we observe that since m_p^* is almost equal to m_n^*, E_i resides close to the center of the band gap. Defining an intrinsic pair product n_i times p_i as

$$n_i p_i = N_c e^{-(E_c - E_i)/kT} N_v e^{-(E_i - E_v)/kT} \tag{34}$$

$$= N_c N_v e^{-E_g/kT} = n_i^2$$

and we obtain the useful conclusion that for *all* n and p at equilibrium

$$np = n_i^2$$

and with Eqs. 28 and 30 any equilibrium carrier concentration can be determined from the position of the Fermi level, or conversely E_F can be found from knowledge of n and p.

7 IMPURITY STATES

Electrons are allowed to occupy energies within the forbidden gap when they are attracted to a potential caused by an impurity or a defect. Electrons attracted to the positive charge of a donor impurity will orbit that positive charge in a way analogous to the orbiting of the nucleus of a hydrogen atom. The donor can therefore be modeled as a hydrogen atom immersed in a dielectric media of relative dielectric constant ϵ_r. The ionization energy of the donor can therefore be calculated from the hydrogenic model:

$$E_B = \frac{m^* q^4}{2\hbar^2 (\epsilon_r)^2 n^2} = \frac{m^*/m}{\epsilon_r^2 n^2} \, 13.6 \text{ eV} \tag{35}$$

Here, m = electron mass in vacuum, n = principle quantum number. Since for compound semiconductors $\epsilon_r \sim 10$ and the effective mass ratio

$m^*/m < 1$, $E_B < 0.1$ eV. The orbit of the electron around the donor can be described from the Bohr radius, a_B:

$$a_B = \frac{\hbar^2 \epsilon_r}{q^2 m^*} = \frac{\epsilon_r}{m^*/m} a_0 \tag{36}$$

where a_0 is the first Bohr radius of hydrogen, equal to 0.53 Å. At low impurity concentrations, these levels reside inside the forbidden energy gap at $\sim E_B$ away from each band edge. Typically, so-called binding energies range from ~ 0.01 to 0.1 eV.

7.1 Impurity Banding

As the impurity concentration approaches $\sim 1/a_B^3$ ($\sim 10^{20}$ cm^{-3}) the electron wavefunctions begin to overlap. Increased wavefunction overlap with increasing doping causes the donor levels to split, and impurity bands form. At such high dopant concentrations, dopants are in close enough proximity that electrons can begin an "impurity band" conduction mechanism by hopping from one donor site to another. The conduction mechanism for the semiconductor at this point becomes metallic-like. Since most donor levels for compound semiconductors are quite shallow, banded impurity levels are therefore likely to enter the conduction band. Electrons traveling within this impurity band are now free to populate conduction band energies. This is the so-called Burstein-Moss effect, illustrated in Fig. 9. Electrons with the appropriate energies are free to populate levels within both the impurity and conduction bands. The ramifications of Burstein-Moss on optical absorption and luminescence are significant, and the effect can be exploited for certain device applications.

7.2 Semi-insulators

Certain impurities, such as transition metals, are capable of multiple ionization and thus can contribute more than one electron or hole to a band. An impurity such as this is called a multiple donor or acceptor. The second ionization energy is usually much higher than the first, and often resides within either the conduction or valence band. Compound semiconductors can be purposely doped with transition metals to make the material semi-insulating. Such semi-insulating material is produced from GaAs doped with Cr or InP doped with Fe. For InP:Fe, the charge from the nominal donor impurity concentration present from unwanted impurities, such as silicon, is compensated by the Fe^{2+}/Fe^{3+} ratio. The ratio is self-adjusting, and changes with the concentration of donors present. The electron, e^-, from a silicon donor in InP which would normally be excited to the conduc-

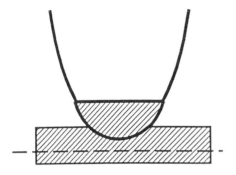

Figure 9 Figure shows impurity banding and the Burstein-Moss effect. The shallow donor levels have banded under high concentration and now electrons are free to populate both the impurity band and the conduction band.

tion band, is instead trapped by the iron impurity:

$$e^- + Fe^{3+} \rightarrow Fe^{2+} \tag{37}$$

The effect is to increase the resistivity of InP to values around 10^7 ohm-cm. Semi-insulating isolation can also be produced in semiconducting material by ion implantation. Implanting B^+, H^+, or oxygen into compound semiconductors creates semi-insulating damage regions in the material useful for isolation. Implantation can then be used to isolate optoelectronic devices from one another.

While semi-insulating substrates are excellent for isolation between devices, and are used extensively in high-speed digital and analog compound semiconductor electronics, their use in optoelectronics is somewhat less common. The semi-insulating wafer is less thermally conductive, and many optoelectronic devices generate heat which must be dissipated.

8   EFFECTS OF STOICHIOMETRY ON CONDUCTIVITY TYPE

For most compound semiconductors, particularly for compounds with a relatively high bonding ionicity, effects of stoichiometry changes during crystal growth or high-temperature processing can alter the conductivity of the material. This is because of the changes made in the chemical activity of native defects present in the material. Changes in conductivity type with stoichiometry are particularly prevalent in II-VI compounds. An example is CdTe, which is p-type with an excess of Cd vacancies and n-type with an excess of Te vacancies [4]. Heating CdTe in excess Te can force changes in the p-type conductivity by forcing reactions such as

$$Te(g) \rightarrow Te_{Te} + V_{Cd} \tag{38}$$

where $Te(g)$ represents the tellurium atom in vapor (or liquid or solid), Te_{Te} represents the Te atom residing on the Te sublattice, and V_{Cd} represents a vacancy on the cadmium sublattice. Because the solid is composed of two different atoms (and then has two different sublattices), a requirement for introducing new atoms in reactions such as Eq. 38, is that the reaction must be site balanced as well as charge and mass balanced. Thus the Te sublattice in the bulk cannot be occupied without the Cd sublattice being "occupied" by an entity of some type, such as a Cd vacancy. Since the missing cadmium atom had a valence of 2, the vacancy is free to "accept" up to two electrons from the valence band. Thus it is capable of being a multiple acceptor:

$$V_{Cd} \rightarrow V_{Cd}^- + h^+ \tag{39}$$

$$V_{Cd}^- \rightarrow V_{Cd}^{2-} + h^+ \tag{40}$$

where h^+ represents a hole in the valence band. The second ionization energy of the vacancy would be much higher than the first, with two electrons then orbiting the vacancy. In a similar fashion, excess Cd creates donor Te vacancies:

$$Cd(g) \rightarrow Cd_{Cd} + V_{Te}^+ + e^- \tag{41}$$

or it may also produce Cd donor interstitials:

$$Cd(g) \rightarrow Cd_i^+ + e^- \tag{42}$$

and the material becomes n-type. In $Hg_xCd_{1-x}Te$, the material becomes

p-type when heated at temperatures above $\sim100°C$, due to the loss of Hg and Cd atoms on cation sites in favor of vacancies and interstitials.

9 EXCITONS AND POLARITONS

9.1 Excitons

An exciton is created from the Coulombic attraction within an electron-hole pair. The result of a free electron and a free hole forming a complex is a lowering of their energy, the creation of a new "entity" (the exciton), and the creation of new energy levels for the exciton to populate. The exciton is free to travel through the crystal, but carries with it no net charge.

The energy levels of the exciton can be calculated by treating the electron as "orbiting" the hole and by using the hydrogenic model:

$$E_{\text{exciton}} = \frac{M_r/m}{\epsilon^2 n^2} \, 13.6 \text{ eV} \tag{43}$$

The mass of the exciton M_r can be determined from calculating a "reduced mass," M_r, for the electron and hole:

$$\frac{1}{M_r} = \frac{1}{m_n^*} + \frac{1}{m_p^*} \tag{44}$$

Because $m_n^* \sim m_p^*$, the exciton mass is small, and the binding energies of excitons are low. The entity is not usually stable at room temperature, and the majority of exciton processes must be observed below 77 K.

The electron and hole that make up the exciton must share the same value of k. The velocities of each of these species are represented individually by

$$v_n = \frac{1}{\hbar} \frac{dE_c}{dk}, \qquad v_p = \frac{1}{\hbar} \frac{dE_v}{dk} \tag{45}$$

which means the exciton only exists at the critical points of the energy bands where

$$\frac{dE_c}{dk} = \frac{dE_v}{dk} \tag{46}$$

Since the exciton is free to move through the crystal, its wavefunction can be modeled as a plane wave:

$$\Psi_{\text{exciton}} = A \exp(i\mathbf{K}_{\text{ex}} \cdot \mathbf{r}) \tag{47}$$

where $\mathbf{r}$ is the position vector with respect to the center of mass of the exciton, and $\mathbf{K}_{ex}$ is the wavevector of the exciton. The introduction of $\Psi_{exciton}$ illustrates that, in terms of how energy is transported, the motion of the exciton in the lattice is very much like that of a free carrier. A major difference is that the exciton transports no net charge and will dissociate under a large enough electric field. The exciton has a range of energies for each $\mathbf{K}_{ex}$, and exciton energy states form "bands" along various crystal directions, very much like those energy bands associated with free carriers.

Due to the exciton's considerably low energy, it is usually only observed at very low temperatures and is easily scattered by defects or impurities. At high doping concentrations, nonuniformities in the doping result in internal electric fields which dissociate the exciton, as shown in Fig. 10a. In a region where the local electric field is caused by a deformation potential, the forces act on the electron and hole in the same direction, and thus the exciton does not dissociate (Fig. 10b). Excitons have important consequences on the optical properties of multiquantum wells and will be discussed again in that context.

9.2 Excitonic Complexes

It is conceivable that three or more particles can form excitonic complexes due to the polar nature of the exciton. An impurity may trap an exciton. In that sense the electron orbits "a neutral acceptor (or hole orbits the

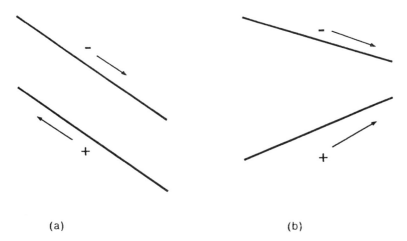

(a)   (b)

Figure 10 Forces exerted on an electron-hole pair (a) under an internal or external electric field and (b) under a deformation potential.

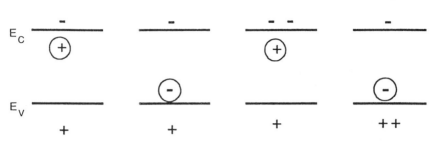

Figure 11 Possible bound exciton complexes. Circled positive (negative) charge indicates the charge of the donor (acceptor) impurity or defect, while the uncircled charges indicate electrons and holes of the exciton.

neutral donor) and the exciton is considered "bound" to the impurity. Figure 11 shows different bound excitons that can be expected to form in the solid. These complexes will further broaden the energy levels of the exciton due to the additional rotational and vibrational modes available.

When the electron and hole at a bound exciton recombine radiatively, the emission is at a photon energy lower than that for the free exciton. The difference is equal to the small binding energy of the bound entity.

9.3 Polaritons

A polariton is a complex between two noncharged entities. The force of attraction between these two entities results from the dipole interaction

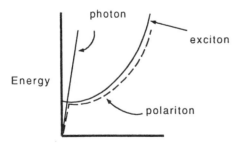

Figure 12 Illustration of how dispersion curves for polaritons are made from that of the free exciton plus the photon. The result—the polariton can only exist in the narrow region where the two curves intersect.

between an electromagnetic wave and an oscillator such as an exciton resonant at the same frequency. Such complexes may include

Excitons and photons
Optical phonons and photons
Plasmons and photons

Phonons are quantized lattice vibrations, and plasmons are quantized free-carrier vibrations. Polaritons formed from photons and excitons can only exist in the narrow region of the intersection of the dispersion curves of the two entities (Fig. 12). The steepness of the dispersion relation for photons severely limits polariton population.

10 DONOR-ACCEPTOR PAIRS

A semiconductor is charge compensated when it contains both positive donor and negative acceptor impurities. There is a Coulombic attraction between the acceptor and donor, which results in the lowering of individual hole and electron binding energies. The strength of the interaction depends on the physical separation between donor and acceptor. Charge compensation increases as the separation between the impurities decreases. The electron is increasingly shared by both impurities as they are brought closer together. When they are at nearest-neighbor sites, the impurities are fully ionized and can no longer donate electrons to the conduction band nor holes to the valence band. The corresponding energy levels for this state then lie at the band edges.

For impurities that are separated such that negligible pairing exists, the donor and acceptor levels are close to the expected hydrogenic energies E_D and E_A. For an intermediate separation distance, r, of the donor-acceptor pair, the energies will be somewhere between the hydrogenic value and the band edge, being shifted by an amount

$$\Delta E = \frac{q^2}{\epsilon r}$$

If both donor and acceptor reside on substitutional sites, r varies in discrete steps. Distinction between donors and acceptors in a compound semiconductor can be made, depending on the atomic sublattice of which they reside. If the donors and acceptors reside on the same sublattice, then they are type I donor-acceptor pair. If they reside on different sublattices then they are type II donor-acceptor pair. An example of type I in GaAs would be silicon on a gallium site as a donor and zinc on a gallium site as an acceptor. Selenium on an arsenic site and zinc on a gallium site would be type II.

11 PERTURBATION OF ENERGY BANDS

A basic understanding of perturbations in semiconductors can be obtained simply by understanding that what causes a localized increase in lattice constant, a, will cause a localized decrease in the energy gap, E_g. Such is true for all semiconductors except PbTe, which has the opposite dependence of E_g with a. A perturbation in a is likely to be caused by pressure or temperature changes. For a change in lattice constant Δa, the resulting perturbation in energy gap ΔE_g is linear as expressed by the relation

$$\Delta E_g = (E_{c1} + E_{v1}) \Delta a \tag{48}$$

where E_{c1} and E_{v1} are linear coefficients determined from

$$E_c = E_{c0} + E_{c1} \Delta a \tag{49}$$

$$E_v = E_{v0} + E_{v1} \Delta a \tag{50}$$

Variations in pressure or temperature and semiconductor alloying are basic processes which result in perturbations in Δa and ΔE_g.

11.1 Band Tailing

If there is localized distortion in the lattice, the result is a localized distortion in the energy bands close to the band edges. Band tail states are formed, which are energy levels extending from the band edges into the forbidden gap, as shown in Fig. 13. For example, band tail states can form from the deformation caused by phonons or from the deformation caused by charged impurities.

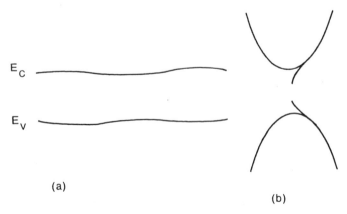

(a)

(b)

Figure 13 Illustration of how localized perturbation of the energy bands (a) produce the band tails in (b).

In a material that contains both donors and acceptors, both positive and negative charges reside in the energy gap. The concentration of these charges builds as the doping is increased. Band tailing is a localized perturbation of the band edges caused by the Coulombic attraction between charged donors and conduction band electrons, and the repulsive forces between the donors and valence band holes. Figure 13 illustrates how the band gap is perturbed along a given direction in a crystal.

11.2 Perturbation under an Electric Field: The Stark Effect

Several of the electron orbits are nonspherical. This includes p orbitals which make up states in the conduction band as well as the elliptical excited states of the exciton. In the bound exciton, if an electron and hole orbit the negative charge of an acceptor impurity, no charge is at the center of the orbit, and the orbit is elliptical. Under an electric field the ellipse aligns itself so the major axis (which contains the negative electron and neutral acceptor) lies along the electric field direction. There will be a shift of the energy level

$$\Delta E = q \, d\mathscr{E} \tag{51}$$

where d is the eccentricity of the elliptical orbit.

11.3 Perturbation Caused by the Franz-Keldysh Effect

An electron wavefunction does not go to zero at a band edge. Instead, a portion of the wavefunction does extend into the band gap. This extension of the wavefunction, which has the form $\sim e^{ikx}$, where k is imaginary, is a damping wave and attenuates with increasing x. Under an electric field the energy levels experience a gradient according to

$$\mathscr{E} = \frac{1}{q}\frac{dE}{dx} \tag{52}$$

and the bands bend. The higher the electric field, the higher the tilt to the bands and the wavefunction then extends farther into the band gap. This is demonstrated in Fig. 14. The probability of finding an electron at an energy E away from the conduction band edge E_c is proportional to

$$\exp\left\{-\frac{E - E_c}{\Delta\mathscr{E}}\right\} \tag{53}$$

where $\Delta\mathscr{E}$ is an electric field parameter given by

$$\Delta\mathscr{E} = \frac{3}{2}(m^*)^{-1/3}(q\hbar\mathscr{E}) \tag{54}$$

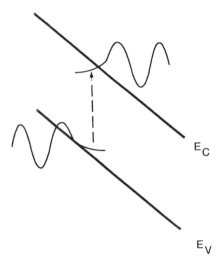

E_C

E_V

Figure 14   Photon-assisted tunneling as per the Franz-Keldysh effect.

The Franz-Keldysh effect is most easily observed during photon-assisted tunneling. As shown in Fig. 14, the extended tails of the wavefunction causes absorption to occur at energies below the band-gap energy, and the absorption edge of the material shifts to longer wavelengths. The higher the electric field, the further the wavefunction extension and, therefore, the greater is the shift.

There is also a two-photon absorption mechanism involving photon-assisted tunneling. A strong electric field shifts the absorption edge to lower energies. The intense electric field of a coherent electromagnetic wave can modulate the absorption edge at optical frequencies. Using two coherent sources of light, both of energy less than the band-gap energy with no field, the absorption of one source can be modulated with the other source. Optical bistability based on effects such as Franz-Keldysh will be discussed in Chapter 6.

11.4   Pressure and Temperature Perturbations

Increasing pressure alters the lattice constant, and hence the critical point energies, according to Eq. 48. For a semiconductor under pressure, the conduction band valleys may move at different rates with increasing pressure. For GaAs under pressure, the direct valley moves to higher energies faster than the indirect, and the band gap, at a critical pressure, becomes indirect. The electron effective mass will also change accordingly.

Temperature effects are slightly more complex than pressure, in that an electron/phonon interaction must also be included. Below the Debye temperature, Θ_D, the energy gap varies as the square of temperature, while above Θ_D, the gap temperature dependence is linear. We may fit the energy-gap temperature dependence to the equation

$$E_g(T) = E_g(0) - \frac{\alpha T^2}{T + \beta} \tag{55}$$

Fitting this equation for curves for GaAs yields values of $E_g(0) = 1.519$ eV, $\alpha = 5.405 \times 10^{-4}$ eV/K, and $\beta = 204$ K [5].

12 OPTICAL ABSORPTION

Many of the optical properties that have been discussed so far can best be understood by observing their affects on the optical absorption. There are many different physical processes involved in absorption in semiconductors, again complicated by the fact that the material is a single crystal and the periodicity of the lattice comes into play in describing electron activity.

When light is incident on a compound semiconductor, a number of processes occur which contribute to reflection, refraction, absorption, and transmission. A hypothetical absorption scan, along with selected critical absorption points, is shown in Fig. 15. Such a scan would be obtained by monitoring the optical transmission through a sample as a function of the wavelength, or energy, of the incident light. Identified in the figure are transitions involving different absorption phenomena. The absorption edge is defined as the transition from valence to conduction band (labeled (3) in the figure). The absorption edge is not a step function, but, because of the existence of band tail states, Franz-Keldysh shifting, and other inhomogeneities in the sample, is instead a exponential increase given by Urbach's rule [6]:

$$\frac{d(\ln(\alpha))}{d(h\nu)} = \frac{1}{kT} \tag{56}$$

where α is the absorption coefficient and $h\nu$ is the energy of the incident photon. The slope of the edge then decreases as temperature increases. The figure also shows intraband absorption (1), interband absorption (3), absorption due to exciton relaxation or donor-acceptor pair complex (2), and absorption due to imperfections, phonons, and free carriers ((4), (5),

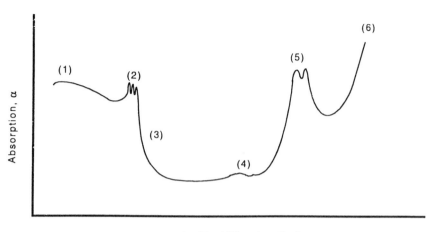

Figure 15 Optical absorption as a function of incident photon wavelength. Reference numbers refer to excitation details as reviewed in the text.

and (6), respectfully). Imperfection absorption is a result of an optically excited transition from a band to an imperfection, or vice versa. An inner-core transition from a neutral to an excited state is also possible for transition metal or rare earth impurities and would give a similar absorption feature to that of (4). Many of the features in Fig. 15, such as the exciton and donor-acceptor pair peaks, are only observable at very low temperatures. Details of so-called resonances (or peaks in the figure) therefore only can be resolved at low temperatures.

For most covalent semiconductors, transitions such as (1) are observable in the UV to visible regions: (2), (3), and (4) in the near- to mid-IR, and (5) and (6) in the mid- to far-IR.

13 COMPOUND SEMICONDUCTOR ALLOYS

Many modern semiconductor optoelectronic devices are formed from compound semiconductor alloys such as $Al_xGa_{1-x}As$. Alloys such as $GaAs_yP_{1-y}$ or $Al_xGa_{1-x}As$ are actually "pseudobinary" alloys of the two individual compounds. For instance, $Al_xGa_{1-x}As$ is the binary alloy of GaAs and AlAs, but these alloys are more commonly referred to as ternary alloys. These alloys must be prepared as single-crystalline films on semiconductor substrates.

Compound semiconductor alloys are prepared by *epitaxy*, a semiconductor crystal growth technique which, although used for many years in silicon technology, has had to undergo radical changes in order to be successful in compound semiconductor applications. In epitaxial crystal growth, thin films of semiconductor alloys are grown on single-crystal substrates. Substrates are primarily, but not limited to, single-crystal wafers of GaAs, InP, or GaP. The film mimics the crystal structure of the substrate on which it sits and therefore has roughly the same crystalline quality. Epitaxy is used in the preparation of optoelectronic devices because it is not possible to prepare pseudobinary and pseudoternary alloys in wafer form.

The three basic types of epitaxy are vapor phase, liquid phase, and molecular beam. The most common techniques for multi-quantum-well (MQW) and superlattice preparation are liquid-phase epitaxy, molecular beam epitaxy, and the gas-phase method known as organometallic vapor-phase epitaxy (OMVPE). There are excellent reviews on these growth techniques, and the reader is referred to these for further information on growth methods [7].

Earlier, the dependence of the energy gap on composition was noted for the alloy Si_xGe_{1-x}. It was found that with increasing x, the indirect and direct valleys move to higher energies at different rates. The rate of change of the energy gap with composition depends on the rate of change of the lowest valley. The same is true for alloys of binary semiconductors, for example the alloy semiconductor $GaAs_yP_{1-y}$ formed from mixing the binary compounds GaAs and GaP. In $GaAs_yP_{1-y}$, the direct gap increases in energy at a higher rate with x than the indirect gap [8]. The indirect and direct gaps are equal when the composition is $x = 0.44$.

The critical points (i.e., the relative placement of energy band edges located at the Γ, L, and X points) are often at different localities for two different compound semiconductors. Mixing the two semiconductors into a pseudobinary, if the mixing is at least somewhat miscible, produces a semiconductor with a type of "engineered" band gap. Because of the chemical differences between the three atoms in a pseudobinary compound semiconductor, the band-gap energy dependence on composition will not be linear.

A quadratic dependence of critical points in the ternary semiconductor ABC formed by alloying binary semiconductors AB plus BC can be modeled as

$$E_{ABC} = xE_{AC} + (1 - x)E_{BC} + x(1 - x)C \tag{57}$$

where E_i represents the critical point energies of compound i. The parameter *C* is known as the bowing parameter. Figure 16 [9] shows values of *C* versus electronegativity differences for several pairs of binary semiconductors used to form ternary semiconductors. The bowing parameter *C* corresponds to the direct-gap energy at the Γ point. We note that it is not necessary for the Γ point to be the smallest energy gap in the pseudobinary.

Fortunately, *Vegard's law* holds for the majority of applications in pseudobinary and pseudoternary alloys. Simply stated, to first order the composition of the alloy is linear with the lattice parameter *a*. Thus, to design an optoelectronic device with a specific band-gap energy, E_g, all one need to know is the dependence of E_g with lattice parameter to determine the required composition.

An important reason to identify the lattice parameter of the alloy is that the alloy must be prepared as an epitaxial film on either a GaAs, InP, or GaP substrate. It must then be "lattice matched" to the substrate. That is, the lattice parameter of the alloy film must be reasonably close to that of the substrate in order to minimize macroscopic defects in the film.

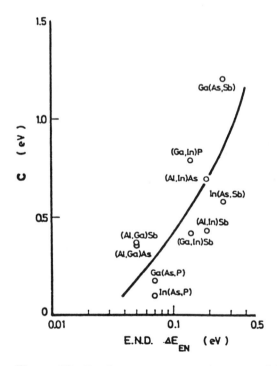

Figure 16 Bowing parameter vs. electronegativity [9].

The compositional dependence of band-gap energies in $Al_xGa_{1-x}As$ also changes from a direct gap to an indirect gap with increasing x. The Γ-X crossover occurs in the composition range $0.4 < x < 0.5$. The energy-gap dependences for the three energy gaps have been quoted by Lee [10] as

$$E_g^\Gamma(x) = 1.425 + 1.155x + 0.37x^2 \tag{58a}$$

$$E_g^X(x) = 1.911 + 0.005x + 0.245x^2 \tag{58b}$$

$$E_g^L(x) = 1.734 + 0.574x + 0.055x^2 \tag{58c}$$

This variation in band-gap energies for the Γ point corresponds to photon wavelengths which vary from the near infrared into the red regions of the spectrum. Because the lattice parameter does not vary appreciably with x for this alloy, it is possible to prepare $Al_xGa_{1-x}As$ on GaAs substrates for all compositions of x. This is not possible for any other III-V alloy system.

If we alloy $GaAs_yP_{1-y}$ with InP (or InAs) we obtain another equally important III-V compound semiconductor, $In_{1-x}Ga_xAs_yP_{1-y}$. This alloy is a pseudoternary alloy, but is more commonly referred to as a quaternary. The quaternary is most useful in near-IR optoelectronics. It is an alloy prepared epitaxially on InP substrates and is lattice-matched to InP for values of x and y such that [11]

$$y = \frac{2.202x}{1 + 0.0659x} \cong 2.2x \tag{59}$$

With this dependence, E_g can be expressed as a function of y only:

$$E_g = 1.35 - 0.72y + 0.12y^2 \text{ eV} \tag{60}$$

This relation was determined experimentally by Nahory et al. [12] using photoluminescence data. A range of energies from 0.75 to 1.65 eV is then possible with the quaternary alloy. This range of energies is important in that the minimum in absorption and material dispersion in silica fiber optics occurs for photon energies within this range (typically ~0.9 eV).

14 QUANTUM WELLS AS OPTICAL MATERIALS

Multiquantum well structures can be prepared from two or more different materials having different band-gap energies. In the previous section, it was noted how the band gap of $Al_xGa_{1-x}As$ increases with increasing x. By growing several thin films of AlGaAs of differing compositions on GaAs substrates, the band gaps can be altered and MQWs prepared. Figure 17 shows a hypothetical quantum well structure. Layers of AlGaAs

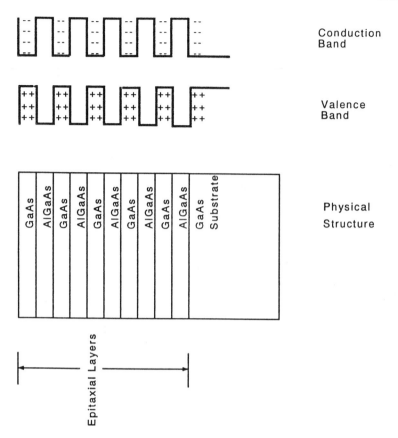

Figure 17 An idealized band structure of a MQW where the electrons occupy the well minima in the conduction band and the holes occupy the barriers in the valence band. The lower figure shows a physical structure of the AlGaAs/GaAs MQW.

and GaAs have been epitaxially deposited in an alternating series on a GaAs substrate. The resulting "idealized" energy band structure is shown in Fig. 17. The periodic steps in the band structure force the electron in the conduction band to reside primarily in the bottom of the energy wells, where their energy is lowest. Similarly, holes in the valence band will rise to the energy barriers which are separating the wells in that band. The barrier represents the minimal energy for the hole, and so the hole will trap there as a particle in a well.

The concept of carriers trapped in energy wells is merely the realization of the particle in the potential well of elementary quantum mechanics. How closely the eigenstates for the structure in Fig. 17 resemble those for the quantum mechanical problem depends upon the width of the well, the barrier height, and the quality of the interface between epitaxial layers.

A more accurate representation of the band structure for multiquantum wells can be determined using a variation of the Bloch functions used in Section 2. We use a wavefunction of the form

$$\psi = \sum_{AB} e^{ik \cdot r} u_k^{A,B}(r) \chi_n(z) \tag{61}$$

where A and B represent the individual films, $u_k^{A,B}(r)$ are the Bloch functions in each A and B layer, and r is a two-dimensional position vector within each layer. The epitaxial growth direction is z. The wavefunction for the three dimensions $u(r_{3D})$ can be determined from

$$u(\mathbf{r}_{3D}) = \Psi(\mathbf{r})\chi(z) \tag{62}$$

and the Hamiltonian is given by

$$H(\mathbf{r}_{3D}) = V_c(z) - \frac{\hbar^2}{2m^*} \frac{\partial^2}{\partial z^2} - \frac{\hbar^2}{2m^*} \frac{\partial^2}{\partial r^2} \tag{63}$$

$\chi_n(z)$ is the envelope function given by

$$\frac{-\hbar^2}{2m^*} \frac{\partial^2 \chi_n(z)}{\partial z^2} + V_c(z)\chi_n(z) = \epsilon_n \chi_n(z) \tag{64}$$

$V_c(z)$ is the energy level at the bottom of the conduction band. The energy state ϵ_n is called the *confinement energy* of the carrier.

The resulting problem is very similar to the common Krönig-Penney model, with the exception of the effective masses. In the multiquantum well, the effective mass "jumps" from one epitaxial layer to the next, which calls for the need of a boundary condition for effective mass:

$$\frac{1}{m_A} \frac{d\chi^A}{dz} = \frac{1}{m_B} \frac{d\chi^B}{dz} \tag{65}$$

In the infinitely-deep-well approximation, the wavefunction goes to zero at the AB interface and the problem reverts to the two-dimensional infinite-well problem from elementary quantum mechanics:

$$\epsilon_n = n^2 \frac{\hbar^2 \pi^2}{2m^* L}, \qquad n = 1,2,3, \ldots \tag{66}$$

with n as an even or odd integer. From the Hamiltonian in the two-dimensional plane

$$H(\mathbf{r}) = -\frac{\hbar^2}{2m^*}\frac{\partial^2}{\partial r^2} \tag{67}$$

we can find solutions for the Schrödinger equation in that plane:

$$H(\mathbf{r})\Psi(\mathbf{r}) = (E - E_n)\Psi(\mathbf{r}) \tag{68}$$

and determine the eigenstates as

$$E_c(k,n) = \frac{\hbar^2 k_\perp^2}{2m^*} + n^2\frac{\hbar^2\pi^2}{2m^*L} \tag{69}$$

where $k_\perp$ is the wavenumber in the two-dimensional plane.

Proof of the quantization of carriers in MQWs was provided by a direct observation of bound-electron and bound-hole states of rectangular potential wells in GaAs/AlGaAs in the optical absorption study of Dingle et al. [13]. The samples were MBE grown superlattices with the substrate selectively removed. The bound states of the carriers within energy wells introduced exciton transitions which could be observed in the absorption spectra.

14.1 Two-Dimensional Density of States

Since the electron, for the most part, is confined to two dimensions, we must look at the probability of occupancy of two-dimensional $k_\perp$-space. The spin-independent density of states (DOS) per unit area can be determined the same way that the three-dimensional DOS is determined. In two-dimensional $k_\perp$-space each point contains two electrons and occupies an area of $(1/2\pi)^2$. A thin circle of area $2\pi k_\perp\, dk_\perp$ is then considered, and the 2D DOS is given by

$$g_{2D}(k_\perp) = 2 \times \left(\frac{1}{2\pi}\right)^2 \times 2\pi k_\perp\, dk_\perp \tag{70}$$

and using the parabolic approximation $E = \hbar^2 k_\perp^2/2m^*$ yields

$$g_{2D} = \frac{m^*}{\pi\hbar^2} \tag{71}$$

The density of states for an energy E_n then is given by

$$g_{2D,n} = \frac{nm^*}{\pi\hbar^2} \tag{72}$$

and increases by a value $m^*/\pi\hbar$ for each increment in energy. The total density of states can be expressed as

$$g_{2D,\text{total}} = \sum_{n}^{\infty} \frac{m^*}{\pi\hbar^2} H(E - E_n) \tag{73}$$

where $H(x)$ is the Heaviside function, which is equal to 1 for $x > 0$ and is 0 when $x < 0$.

The density of states then, is independent of energy and layer thickness. Compared to the three-dimensional density of states

$$g_{3D} = 2^{1/2} m^{*3/2} \pi^{-2} h^{-3} E^{1/2} \tag{74}$$

we note that, for the 2D DOS, g_{2D} does not go to zero at the band edges. For 3D conduction, the probability of certain phenomenon such as scattering, optical absorption, and optical gain approaches zero at low temperatures, as the interactions involve free carriers closer to the band edges. This is due to the zero DOS at the band edges of three-dimensional conduction materials. For two-dimensional materials, there is a finite probability of observing these phenomena at any temperature. When many levels in the bands are populated, such as the case for highly excited semiconductor lasers, the 2D DOS appears parabolic as the 3D.

14.2   Bound Excitons in MQWs

In Section 9.1 we observed how the Coulombic attraction between free electrons and holes can result in a exciton which is free to travel through the crystal. It was also shown how this entity could bind itself to charged impurities. In a 2D multi-quantum well the entity can be described using a hydrogenic model, in which it has an effective radius given by the Bohr formula

$$a_B = \frac{4\pi\epsilon_0\epsilon_r\hbar^2}{M_{\|}q^2} \tag{75}$$

where $M_{\|}$ is the reduced mass of the electron as given in Eq. 44, only with 3D values of the effective mass substituted with 2D values. In the limiting case, the well spacing actually becomes smaller than the Bohr diameter $2a_B$. Although the Coulombic potential is still 3D and goes as $\sim 1/r$, the major effect of the restriction of dimensionality is to reduce the dimensionality of the exciton orbits. The wavefunction now has no dependence on azimuth angle θ, the main effect being the principle quantum number n becomes $n - 1/2$. The spacing between the energy levels of the normal 3D Rydberg series R_{3D} are increased in the 2D series and R_{2D} is given by

$$R_{2D} = \frac{R_{3D}}{(n - 1/2)^2} \quad \text{where } R_{3D} = \frac{q^4 M_{par}}{\epsilon^2 \hbar^2} \tag{76}$$

The exact binding energy for the exciton will depend on several variables. In a multilayer system, as the layer thickness decreases, there will be an increase in overlap of the exponential tails of the carriers which extend into the wide-band-gap layer. This overlap is large enough to give 3D character to the system. The exciton binding energy then decreases toward the 3D value. For layers that are relatively thick, the 2D confinement decreases, and again the exciton energy decreases to the 3D value. There is, therefore, an optimum thickness to obtain a maximum binding energy.

As the exciton binding energy is increased above the 3D value, it becomes theoretically possible to observe excitons at a higher temperature in MQW systems. Such high-temperature excitons are of technologic importance to devices which depend on exciton concentration, such as certain self-electro-optic effect devices (SEED) discussed in Chapter 6.

REFERENCES

1. W. A. Harrison, *Solid State Theory*, Dover, New York, 1979.
2. J. R. Chelikowsky and M. L. Cohen, Nonlocal psuedopotential calculations for the electronic structure of eleven diamond and zinc-blende semiconductors, *Phys. Rev. B, 14*, 556 (1976).
3. J. I. Pankove, *Optical Processes in Semiconductors*, Prentice Hall, Englewood Cliffs, NJ, 1971, p 18.
4. F. A. Kröger, *The Chemistry of Imperfect Crystals*, Vol. 2, 2nd ed., North-Holland, Amsterdam, 1974, p. 728.
5. C. D. Thurmond, The standard thermodynamic functions for the formation of electrons and holes in Ge, Si, GaAs, and GaP, *J. Electrochem. Soc., 122*, 1133 (1975).
6. F. Urbach, The long wavelength edge of photographic sensitivity and the electronic absorption of solids, *Phys. Rev., 92*, 1324 (1953).
7. S. K. Ghandhi, *VLSI Fabrication Principles: Silicon and GaAs*, Wiley, New York, 1983, Chapter 5.
8. J. J. Tietjen, and J. A. Amick, The preparation and properties of vapor-deposited eptaxial $GaAs_{1-x}P_x$ using arsine and phosphine, *J. Electrochem Soc., 113*, 724 (1966).
9. S. Adachi, GaAs, AlAs, and $Al_x Ga_{1-x}As$: material parameters for use in research and device applications, *J. Appl. Phys., 58*, R1–R29 (1985).
10. H. J. Lee, L. Y. Juravel, J. C. Wooley, and A. J. Springthorpe, Electron transport and band structure of $Ga_{1-x}Al_xAs$ alloys, *Phys. Rev. B, 21*, 659 (1980).

11. T. P. Pearsall, Electronic structure of $Ga_xIn_{1-x}As_yP_{1-y}$ alloys lattice-matched to InP, in *GaInAsP Alloy Semiconductors* (T. P. Pearsall, ed.), Wiley, New York, 1982, p. 295.
12. R. E. Nahory, M. E. Pollack, W. D. Johnston, and R. L. Barns, Bandgap versus composition and demonstration of Vegard's law for $In_xGa_{1-x}As_yP_{1-y}$ lattice matched to InP, *Appl. Phys. Lett.*, *33*, 659 (1978).
13. R. Dingle, W. Weigmann, and C. H. Henry, Quantum states of confined carriers in very thin $Al_xGa_{1-x}As–GaAs–Al_xGa_{1-x}As$ heterostructures, *Phys. Rev. Lett.*, *33*, 827 (1974).

BIBLIOGRAPHY

1. J. I. Pankove, *Optical Processes in Semiconductors*, Prentice Hall, Englewood Cliffs, NJ, 1971.
2. R. H. Bube, *Electronic Properties of Crystalline Solids*, Academic Press, New York, 1974.
3. T. S. Moss and M. Balkanski, (eds.), Optical properties of solids, in *Handbook on Semiconductors*, Vol. 2, North-Holland, Amsterdam, 1980.

2
Light-Emitting Diodes

1 INTRODUCTION

Light-emitting diodes (LEDs) have become one of the most successful optoelectronic components of the last 30 years. Even though the LED may not be as technologically glamorous as its relative the laser diode, the latter is more difficult to fabricate, requires more complex drive and cooling circuitry, and is overall more complex than the LED. LEDs are not subject to certain failure mechanisms that plague lasers, such as variations in threshold current or facet damage.

Even though many applications of LEDs as display devices have been supplanted by liquid crystal technology, red, yellow, green, and blue LEDs are still a major commercial success. By far, the most technologically interesting application, however, has been infrared LEDs for optical communications. Such devices have achieved excellence as optical sources for fiber optics and thin-film waveguides.

A light-emitting diode could be defined as a p-n junction under forward bias undergoing radiative recombination. The abrupt junction formed by an n-type and a p-type semiconductor when brought together has an altered band structure as shown in Fig. 1. Under forward bias, majority carriers (electrons and holes) from n- and p-type regions are injected across the junction into the region of opposing conductivity, where they instantly become minority carriers. As minority carriers they diffuse a

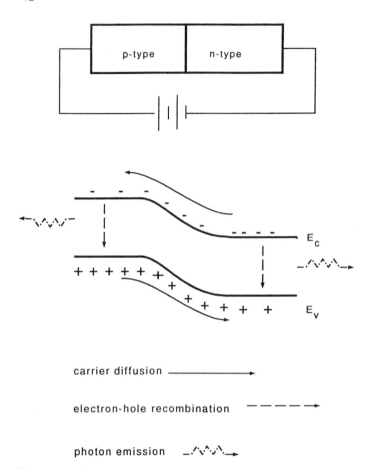

Figure 1 The process of carrier injection in a forward-bias diode. The electrons and holes diffuse across the boundary of the *p-n* junction where they become minority carriers. Here they recombine, some emit radiation, and some do not.

small distance in the region and eventually recombine with majority carriers. As carriers recombine they lose energy which, if conditions permit, is emitted as photons. Recombination that produces photons is termed *radiative*, all other types being nonradiative. The success of LED operation depends on several parameters, the majority of which are ultimately fabrication dependent. Steps must be taken to insure that a large portion of the recombination energy is emitted as electromagnetic rather than thermal energy. As heat is invariably generated, it must be efficiently dissipated. Once photons are generated, the probability of the light successfully exiting the device may be quite small, so that an appreciable amount of radiative recombination is required.

We divide this chapter into three parts. First, physics of the *p-n* junction is reviewed. Then the mechanisms of recombination, and then the structures and materials of the devices are studied.

2 REVIEW OF *p-n* JUNCTION PHYSICS

2.1 The Continuity Equation

In order to understand the operations of conduction and subsequent photon emission in a light-emitting diode, we need to first analyze the basic conduction mechanisms of a semiconductor. We will need, in order to obtain efficient radiative recombination, a concentration of *excess* carriers in one or each side of the *p-n* junction. That is, we require a concentration of minority carriers above the equilibrium value in order to create a driving mechanism for recombination. At the same time, a current must flow to maintain this excess concentration. The analysis of the mechanisms of current flow are therefore important for proper device operation.

We will use as our example electron flow in the conduction band. The density of electrons in this band will change as a result of several processes including (1) generative processes, or more simply *generation*, where electrons are generated from some lower energy state, either from the valence band or the from a level within the band gap. (cf. Fig. 2); (2) *recombination*, which can be viewed as the opposite of generation, and can be either radiative or nonradiative in nature; (3) *drift and diffusion*, both of which are considered the more basic current flow mechanisms in a semiconductor.

Generative processes include any mechanism which would transfer enough energy to an electron as to cause it to cross from the valence band to the conduction band. A direct band-to-band generation of an electron in the conduction band results in the creation of a hole in the valence band (Fig. 2). When a new electron-hole pair is created by a band-to-band transition, an energy of $\sim(E_c - E_v)$ is necessary to complete the process. However, electrons (and holes as well) can be generated from levels within the forbidden gap. This includes reexcitation of electrons that may previously have been trapped at these energy levels. Different generative processes are illustrated in Fig. 2.

In addressing details of carrier generation and recombination it is customary to separate "external" from "internal" generative processes, However, the exact definition of each varies in the literature. The rate of

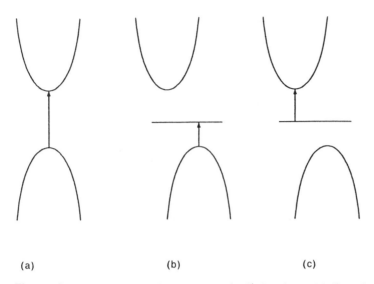

(a)   (b)   (c)

Figure 2   Possible generation processes in the band gap: (a) direct band to band, resulting in the creation of an electron-hole pair. (b) Band to level, and (c) level-to-band. The latter two processes are illustrated in indirect-gap materials merely for example.

internal generation, g, is sometimes loosely defined as the rate of all *thermally* generated excitation processes, and external includes all else. Technically speaking, however, g also includes any electron-hole generation produced by normal blackbody background electromagnetic radiation. For a wide-gap material this component may be small enough to be neglected, and only thermal processes considered. The rate of external generation, g_{ex}, includes any excitation caused by excess electromagnetic radiation generated by radiative recombination in the forward-bias diode or by external stimuli such as cosmic rays. The important distinction is that the internal generative processes include *normal* processes which would occur if no current was flowing and the device was not exposed to outside influences.

The total rate of generation is then $g + g_{ex}$. Likewise, the total rate of recombination is then $r + r_{ex}$, where r_{ex} includes recombination *induced* by excess electromagnetic radiation. We will use g_E to denote the net external rate $g_{ex} - r_{ex}$.

Current flow can also change the electron density in the conduction band. We consider a volume of semiconductor of size $dx \cdot dy \cdot dz$ and analyze the current as it flows in and out of each side of that volume. In the x direction, the rate at which electrons flow into the element due to the current flux J_{nx} is

$$-\frac{J_{nx}}{q} \cdot dy \cdot dz \tag{1}$$

while the flow out of the element in the x direction is given as

$$\left(-\frac{J_{nx}}{q} - \frac{1}{q}\frac{\partial J_{nx}}{\partial x} \, dx\right) dy \cdot dz \tag{2}$$

and the net rate of electron accumulation in the volume is

$$\frac{1}{q}\left(\frac{\partial J_{nx}}{\partial x}\right) dx \cdot dy \cdot dz \tag{3}$$

We can use the same technique to find the net rate of electron current flow into the other two faces of the volume:

$$\frac{1}{q}\left(\frac{\partial J_{nx}}{\partial x} + \frac{\partial J_{ny}}{\partial y} + \frac{\partial J_{nz}}{\partial z}\right) dx \cdot dy \cdot dz \tag{4}$$

$$= \frac{1}{q}\nabla \cdot \mathbf{J}_n \, dx \, dy \, dz \tag{5}$$

The rate of change of local free electron density (electrons per volume element $dx \, dy \, dz$), taking into account generation and recombination as mentioned above, is

$$\frac{dn}{dt} = (g - r) + g_E + \frac{1}{q}\nabla \cdot \mathbf{J}_n \tag{6}$$

The equation for hole accretion is

$$\frac{dp}{dt} = (g' - r') + g'_E - \frac{1}{q}\nabla \cdot \mathbf{J}_p \tag{7}$$

The generation/recombination terms g, g_E, and r for electrons are equal to g', g'_E, and r' for holes only when band-to-band processes dominate the recombination.

We can replace $g - r$ in Eq. 6 with $(n - n_0)/\tau_n$, where τ_n is the bulk carrier lifetime and n_0 is the equilibrium carrier concentration. τ_n is either completely independent or only weakly dependent on $n - n_0$.

There should be some discussion on what is defined as "recombination rate." In many texts, a recombination rate U is defined which is actually the *net* recombination rate, $r - g$. Sometimes $1/\tau_n$ is also incorrectly defined as the recombination rate or recombination rate constant. We will be aware that recombination rate r is sometimes synonymous with net recombination rate $r - g$.

2.2 *p-n* Junction

The LED is considered a minority carrier device—that is, a device where minority carrier injection and diffusion play an important role in determining the device characteristics. The minority carriers must be injected, diffuse a distance in the opposing conductivity region, then recombine with majority carriers.

There are two primary mechanisms for current transport in semiconductors: diffusion and drift. Diffusion is motion of carriers under a concentration gradient; in one dimension, dn/dx or $dp/dx \neq 0$. Drift is carrier motion under an electric field. The overall current is the sum of all diffusion plus drift currents.

$$J_{\text{total}} = J_{\text{diffusion}} + J_{\text{drift}}$$

where J is the current flux. Total diffusion currents in a semiconductor are the sum of p and n carrier diffusion currents:

$$J_{\text{diffusion}} = J_{p\ \text{diffusion}} + J_{n\ \text{diffusion}}$$

and a similar expression can be written for drift currents. Total hole currents, as the sum $J_{p\ \text{drift}}$ plus $J_{p\ \text{diffusion}}$, can be represented by the well-known equation:

$$\mathbf{J}_p = q\mu_p p\mathcal{E} + qD_p\nabla p \tag{8}$$

where $\mathcal{E}$ is the electric field, ∇p is the hole concentration gradient, and D_p is the diffusion coefficient for holes.

When neutral n- and p-type materials are brought together, the high electron concentration in the n-type side is offset by a very low electron concentration on the p-type side, and the converse is true for holes. Such high disparity in carrier concentration is a driving force for electron and hole diffusion currents. When the carriers diffuse, the fixed charge of the n-type donors and the p-type acceptors is exposed, and an internal electric field is generated (Fig. 3). This electric field is in a direction that opposes diffusion flow, and a drift current is set up that, at equilibrium, balances the diffusion current. At equilibrium J_p in Eq. 8 $= 0$ and we may write

$$q\mu_p p\mathcal{E} = qD_p\frac{dp}{dx} \tag{9}$$

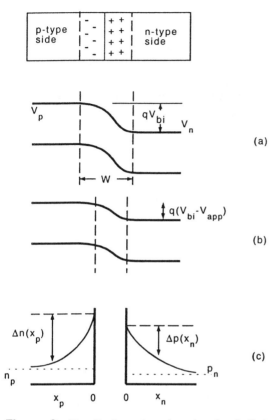

Figure 3 Detail of *p-n* junction showing built-in field qV_{bi} and the depletion width W (a) with no bias, (b) with forward bias. (c) Minority carrier concentration as a function of distance injected from the junction.

If we substitute for the electric field gradient and the Nernst-Einstein relation, respectively, then

$$\mathcal{E} = -\frac{dV}{dx}, \quad \frac{D_p}{\mu_p} = \frac{kT}{q} \tag{10}$$

We can obtain the expression

$$-\int_{V_p}^{V_n} dV = \frac{kT}{q} \int_{P_p}^{P_n} \frac{1}{p} \, dp \tag{11}$$

In the left side of Eq. 11 we are integrating from the potential on the n-type side to that on the p-type side. The reference may be taken at any place on the energy bands, for example at E_c or E_v. Similarly, we integrate the right side of the equation from the n-type side (where the holes are minority carriers) to the p-type side. In each case p_p and p_n are taken as the equilibrium or bulk concentrations. Defining the built-in potential as $V_{bi} = V_n - V_p$ we have, on integration of Eq. 11,

$$V_{bi} = \frac{kT}{q} \ln\frac{p_p}{p_n} \tag{12}$$

If we have a concentration N_A of acceptors in the p-type side and N_D donors on the n-type side, and the temperature is high enough that all donors and acceptors are ionized (which is usually the case in compound semiconductors), we can write $N_A = p_p$ and $N_D = n_n$. Along with the useful equation, $n_n = n_i^2/p_n$, we obtain an equation for the built-in potential:

$$V_{bi} = \frac{kT}{q} \ln\frac{N_A N_D}{n_i^2} \tag{13}$$

Under the forward bias shown in Fig. 3b, the applied electric field is opposite to the built-in field, and the drift component of the current is reduced. The height of the barrier is reduced, diffusion is then free to take over, and carriers are "injected" into the opposite conductivity type. Here they diffuse a short distance determined by the minority carrier lifetime until they recombine.

2.3 The Depletion Width

For a sample of cross section A, the sum of the positive charges on one side of the semiconductor junction must equal the sum of the negative charges on the other side. If x_n and x_p are, respectfully, the widths of the depletion region into the n-type and the p-type sides, for charge compensation,

$$N_A x_{p0} = N_D x_{n0} \tag{14}$$

the total width of the depletion region $W = x_{n0} + x_{p0}$ can be found from Poisson's equation:

$$\frac{d^2V_n}{dx^2} = \frac{qN_D}{\epsilon_s} \quad \text{and} \quad \frac{d^2V_p}{dx^2} = \frac{qN_A}{\epsilon_s} \tag{15}$$

Integrating each equation twice we have

$$V_n = \frac{qN_D}{2\epsilon_s} x_{n0}^2$$

$$V_p = \frac{qN_A}{2\epsilon_s} x_{p0}^2$$

(16)

Since $V_{bi} = V_n - V_p$ and $W = x_{n0} + x_{p0}$, we have a relation for W:

$$W = \left[\frac{2\epsilon_s V_{bi}}{q}\left(\frac{1}{N_A} + \frac{1}{N_D}\right)\right]^{1/2}$$

(17)

For a one-sided abrupt junction, the doping is heavier on one side of the junction, and the depletion width extends into the low-doped side. Using a p^+n junction as an example (p^+ = heavy acceptor doping), we get

$$W = \left[\frac{2\epsilon_s V_{bi}}{qN_A}\right]^{1/2}$$

(18)

$$= L_D\left(\frac{2qV_{bi}}{kT}\right)^{1/2}$$

(19)

where L_D is the Debye length, defined as

$$L_D = \left(\frac{\epsilon_s kT}{q^2 N_A}\right)^{1/2}$$

(20)

For GaAs abrupt junctions at equilibrium, the depletion layer widths are on the order of $10L_D$. From these results we see that the depletion width decreases with increasing doping of the low-doped side.

The depletion layer width under an applied voltage can be written as

$$W = L_D\left(\frac{2q(V_{bi} \pm V_{app})}{kT}\right)^{1/2}$$

(21)

where $V_{bi} + V_{app}$ is used for an applied reverse bias and $V_b - V_{app}$ is used for forward. For forward bias the sign is opposite that of V_{bi} since the direction of the field is opposite to that of the built-in field.

The depletion layer capacitance per unit area can be found from

$$C = \frac{\epsilon_s}{W}$$

(22)

Both the width and capacitance depend on the square root of the applied field. In reverse bias, increasing the field increases the width and decreases the capacitance. Under forward bias the opposite holds. Under a large forward bias, in the *heavy injection* condition, there will be an additional component of the capacitance due to the charge of the excess carriers present. It is customary to write depletion-layer capacitance as

$$\frac{1}{C^2} = \frac{2L_D^2}{\epsilon_s^2} \frac{q(V_{bi} \pm V_{app})}{kT} \tag{23}$$

By plotting $1/C^2$ versus V_{app} for an abrupt, one-sided junction the result is a straight line, and one can determine the doping density N_A from the slope.

2.4 Excess Carriers

Figure 3c shows the variation of minority carrier concentration as a function of distance away from the junction as the carriers are injected in forward bias. The excess carrier concentration on each side is given by $\Delta p(x_n)$ for the hole concentration on the n-type side and $\Delta n(x_p)$ for the corresponding concentration of electrons on the p-type side. The concentrations are measured for two different coordinate systems, x_n and x_p, with origin for each at the edge of the depletion region. The equilibrium minority carrier concentration in the p-type region $n(x_p = \infty) = n_p$. From Eq. 12 we have

$$\frac{n_n}{n_p} = e^{qV_{bi}/kT} \tag{24}$$

With an applied forward bias V_{app} we can rewrite this equation as

$$\frac{n_n}{n(x_p = 0)} = e^{q(V_{bi} - V_{app})/kT} \tag{25}$$

We first make the approximation that the majority carrier concentration does not vary appreciably under injection, i.e., $n_n(x_n) = n_n$ which is constant. This is the so-called *low injection* condition. Writing $n(x_p = 0)$ as $n(0)$ we have

$$\frac{n(0)}{n_p} = e^{qV_{app}/kT} \tag{26}$$

By definition

$$\Delta n(0) = n(0) - n_p = n_p(e^{qV_{app}/kT} - 1) \tag{27}$$

and for the n-type side

$$\Delta p(0) = p_n(e^{qV_{app}/kT} - 1) \tag{28}$$

We can describe the exponential decay from the point of injection ($x_p = 0$) as

$$\Delta n(x_p) = \Delta n(0)e^{-x_p/L_n} \tag{29}$$

where L_n is the minority carrier diffusion length, which is related to the minority carrier lifetime τ_n and diffusion coefficient D_n by

$$L_n = \sqrt{D_n \tau_n} \tag{30}$$

Substituting for $\Delta n(0)$ we obtain an equation for the decay of minority carriers in the p-region:

$$\Delta n(x_p) = n_p(e^{qV_{app}/kT} - 1)e^{-x_p/L_n} \tag{31}$$

Since the currents are all diffusion currents in low-level injection, we can write the total current flux J as the sum of electron current flux J_n and hole current flux J_p:

$$J = J_n + J_p \tag{32}$$

We use the equation for diffusion currents to describe each component:

$$J_n = -qD_n \frac{d(\Delta n(x_p))}{dx_p}$$
$$J_p = -qD_p \frac{d(\Delta p(x_n))}{dx_n} \tag{33}$$

Although carrier flux is in opposing directions, the signs of J_n and J_p are the same in our case since x_p is positive in a direction opposite to x_n. Since all the injected carriers must pass through the points $x_p = 0$ for electrons and $x_n = 0$ for holes, we can evaluate each component of the current at these points:

$$J_n = J_n(0) = q\frac{D_n}{L_n}\Delta n(0)e^{-x_p/L_n}\bigg|_{x_p=0} \tag{34}$$

$$= q\frac{D_n}{L_n}\Delta n(x_p)\bigg|_{x_p=0} \tag{35}$$

$$= q\frac{D_n}{L_n}n_p(e^{qV_{app}/kT} - 1) \tag{36}$$

Similarly, for hole diffusion currents,

$$J_p = q\frac{D_p}{L_p}p_n(e^{qV_{app}/kT} - 1) \tag{37}$$

where p_n is the equilibrium minority carrier concentration in the n-side. Writing $J = J_n + J_p$ gives us the diode equation

$$J = J_0(e^{qV_{app}/nkT} - 1) \qquad (38)$$

where J_0 is known as the reverse saturation current and is defined as

$$J_0 = q\left(\frac{D_n}{L_n}n_p + \frac{D_p}{L_p}p_n\right) \qquad (39)$$

In Eq. 38 under reverse bias the sign of V_{app} is negative and $J = J_0$.

In the diode equation the parameter n is known as the ideality factor and takes into account the finite resistance of the neutral n- and p-regions of the device. The ideality factor usually varies from 1 to 2.

After the minority carriers have been injected, they diffuse a few diffusion lengths and then recombine. Recombination then forces a change in the majority carrier concentration, and the remainder of the current is majority carrier current. The electrodes attached to the device must resupply the majority carriers lost through recombination.

3 RECOMBINATION

A semiconductor contains excess majority and minority carriers when these carriers are either injected from a forward-biased p-n junction or when they are excited by external radiation. Such external radiation may be from a beam of photons or high-energy electrons. When a semiconductor is subjected to excess carriers such that the np product suddenly becomes greater than n_i^2, there will be a tendency for that semiconductor to try to restore itself to $np = n_i^2$, through recombination, i.e., through the reaction

$$e^- + h^+ \rightarrow \text{null} \qquad (40)$$

where e^- represents a conduction band electron, h^+ a valence band hole, and null represents the equilibrium condition. Figure 4 shows that reactions such as (40) may proceed directly across the band gap from conduction to valence band (band-to-band recombination) or through a defect center; that is, an energy level within the forbidden gap. There will be energy ($\sim E_g$) released in Reaction 40, and that energy may be released in various forms or transferred to various entities. In a direct-gap semiconductor, the band-to-band transition is vertical, and a large part of the recombination energy is released as photons. For an indirect band-to-band transition (Fig. 4b) one or more phonons are necessary to provide momentum conservation to the recombination, and the energy gained

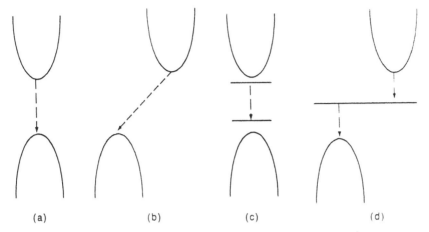

Figure 4 Recombination mechanisms: (a, b) band to band, for direct- and indirect-band-gap materials, respectively; (c) donor-acceptor pair recombination, (d) recombination through a center in the energy gap.

through recombination will be released to these phonons. Because of the large number of phonons that would be involved, this is an unlikely event, and recombination is not probable in indirect materials that do not contain a recombination center (such as undoped silicon).

Energy can also be given to an extra electron, if one is available, in *Auger recombination*, and this is also considered "radiationless" recombination.

It is most probable that nonradiative recombination will take place through a recombination center. The carrier would first be captured at an excited center, then follow successive decreases in energy down to the ground state with multiple-phonon emission. The center is also free to absorb some of this energy through localized vibrations of its own. The result is the major source of heat during LED operation.

When recombination involves a defect center, the center may be termed a "recombination" or a "trapping" center, depending upon the length of time it spends in that center. Recombination through a center can be distinguished from trapping, where the electron spends time and then is rejected back into the conduction band:

$$e^- + X_t^+ \rightarrow X_t^0 \rightarrow X_t^+ + e^- \tag{41}$$

where X_t^+ represents a *trapping center*, which is a defect with an energy state in the energy gap where the carrier temporarily resides. The defect could also act as a *recombination center* (Fig. 4d). Electron-hole recombination through a center means an intermediate step must be added to Eq. 40:

$$e^- + X_t^+ \rightarrow X_t^0 \tag{42a}$$

$$X_t^0 + h^+ \rightarrow X_t^+ \tag{42b}$$

Designating the recombination center as having a positive charge is only used here for illustrative purposes. The recombination center could be neutral:

$$e^- + X_t^0 \rightarrow X_t^- \qquad X_t^- + h^+ \rightarrow X_t^0 \tag{43}$$

The recombination center could be negative as well. We note that the recombination center regenerates itself, so it may recombine more carriers. It is also possible that the hole may be captured before the electron. For a select few recombination centers, either electron or hole capture (Eq. 42a or Eq. 42b, for example) may be a radiative processes.

The motivation to restore the semiconductor back to equilibrium is proportional to the *excess* carrier concentration(s), $\Delta n = n - n_0$, and $\Delta p = p - p_0$. Note that either electron, or hole, or both concentrations may be in excess, depending on the generation process. The *carrier lifetime* τ_n may be loosely defined as the lifetime of the electron in the band, not including the time spent in a trap. Radiative, nonradiative, or Auger recombination may result whether recombination is band to band or takes place through a recombination center.

Band-to-band recombination energies may include the relaxation energies of several entities mentioned in Chapter 1. Free and bound excitons, polaritons, and donor-acceptor pair entities all may be present in an excited semiconductor under the right conditions. Recombination energy can be transferred to and from any of these entities.

3.1   Radiative Recombination

Radiative recombination can be modeled using the van Roosbroeck–Shockley relations. This analysis is based on deriving an equation that describes generation in semiconductors, and then using the principle of detailed balance to derive a recombination rate. The principle of detailed balance simply states that at equilibrium the transition probabilities for generation and recombination are equal.

At equilibrium the rate of electron-hole generation ($g_0(\nu)$) by light of frequency ν is given by

$$g_0(\nu) \, d\nu = P(\nu)\rho(\nu) \, d\nu \tag{44}$$

where $\rho(\nu)\, d\nu$ is the density of photons in a frequency interval $d\nu$. $P(\nu)$ is the probability per unit time of absorbing a photon of energy $h\nu$, which is given by

$$P(\nu) = \frac{1}{\tau(\nu)} \tag{45}$$

where $\tau(\nu)$ is the mean lifetime of the *photon* in the semiconductor. The absorption length is $1/\alpha$, where α is the absorption coefficient. For a photon of velocity $v = c/n$, the photon lifetime, therefore, is

$$\tau = \frac{1}{\alpha v} = \frac{n}{\alpha c} \tag{46}$$

which gives

$$P(\nu) = \frac{\alpha c}{n} \tag{47}$$

The quantity $\rho(\nu)$ can be obtained from Plank's radiation law for a black body source:

$$\rho(\nu)\, d\nu = \frac{8\pi\nu^2 n^3}{c^3} \frac{1}{\exp(h\nu/kT - 1)}\, d\nu \tag{48}$$

The rate per unit volume at which electron-hole pairs are generated is then given by

$$g_0(\nu) = \int \frac{\alpha c}{n} \rho(\nu)\, d\nu \tag{49}$$

$$= \frac{8\pi n^2}{c^2 h^3} \int \frac{\alpha(h\nu)^2\, d(h\nu)}{\exp(h\nu/kT - 1)} \tag{50}$$

The greatest contribution to the above integral will be at the absorption edge. The principle of detailed balance suggests that since the recombination rate is equal to the generation rate at equilibrium, we may write $r_0(\nu) = g_0(\nu)$, and the recombination rate is also given by the integral in Eq. 50. When E_g is large compared with kT we can express the denominator in the integral simply as $\exp(h\nu/kT)$ and the integral can be determined numerically [1]. The recombination rate can now be expressed as

$$r_0(\nu) = \frac{8\pi n^2}{c^2 h^3} \int \frac{\alpha(h\nu)^2\, d(h\nu)}{\exp(h\nu/kT)} \tag{51}$$

Expressions can now be obtained for the rates of spontaneous and stimulated recombination. The variation of the generation rate per energy increment dE for a band-to-band transition can be expressed by

$$\frac{dg_0(\nu)}{dE} = A P_n(E_v) P_p(E_c) \rho(\nu) \tag{52}$$

where $P_n(E)$ ($P_p(E)$) is the probability that a state is occupied by an electron (hole). The electron is making a transition from *some* state E_v in the valence band to a state E_c in the conduction band. The proportionality constant, A, depends on the density of states and the form of the wavefunction for electrons in each band. At equilibrium $P_n(E)$ is equal to the Fermi function:

$$F(E) = \frac{1}{\exp(E - E_F)/kT} \tag{53}$$

Under nonequilibrium conditions, $P_n(E)$ can be represented by this equation if the equilibrium Fermi energy E_F is replaced by the appropriate quasi-Fermi energy. Since $P_n(E) = 1 - P_p(E)$, it can easily be shown that

$$[P_n(E_v) P_p(E_c)] = \exp(E_c - E_v) P_n(E_c) P_p(E_v) \tag{54}$$

The total recombination rate is the sum of spontaneous (r_{sp}) plus stimulated (r_{st}) emission:

$$\frac{dr_0(h\nu)}{dE} = \frac{dr_{\mathrm{sp}}}{dE} + \frac{dr_{\mathrm{st}}}{dE} \tag{55}$$

$$= P_n(E_c) P_p(E_v) B + A' P_n(E_c) P_p(E_v) \rho(\nu)$$

where B is the spontaneous recombination constant and A' is the recombination constant for stimulated emission. At equilibrium, Eqs. 52 and 55 are equal, and assuming $A = A'$ we obtain

$$B = A\rho(\nu) \left[\exp\left(\frac{h\nu}{kT}\right) - 1 \right] \tag{56}$$

Substituting in for $\rho(h\nu)$ from Eq. 48 gives a relation between the stimulated and spontaneous emission constants:

$$\frac{B}{A} = 8\pi\nu^2 \frac{n^3}{c^3} \tag{57}$$

We can also determine a relationship between spontaneous and stimulated emission:

$$\frac{dr_{\mathrm{sp}}}{dr_{\mathrm{st}}} = \exp\left(\frac{h\nu}{kT}\right) - 1 \tag{58}$$

We can now write expressions for spontaneous and stimulated emission from Eq. 51:

$$r_{sp}(\nu) = \frac{8\pi n^2}{c^2 h^3} \int \frac{\alpha(h\nu)^2 \, d(h\nu)}{\exp(h\nu/kT)}$$

$$r_{st}(\nu) = \frac{8\pi n^2}{c^2 h^3} \int \frac{\alpha(h\nu)^2 \, d(h\nu)}{\exp(h\nu/kT)[\exp(h\nu/kT - 1)]}$$

(59)

These equations hold for band-to-band as well as recombination through a center. Using Eq. 59, one can take a known absorption spectrum and generate an emission spectrum. This has been done to calculate such recombination parameters as recombination constants and lifetimes, as discussed in the next section. It is evident from Eq. 59 that low values of the absorption coefficient translate to small recombination rates and long carrier lifetimes. An indirect-gap semiconductor with a weak absorption edge will make a poor laser.

Since there are nonequilibrium carrier densities in a forward-biased LED, emission rates will depend on the level of carrier injection and will be different from Eq. 59. If the semiconductors are not doped to degeneracy, and lifetimes are relatively long, the recombination rate for spontaneous emission under nonequilibrium conditions can be written as

$$r = \frac{np}{n_i^2} r_0 = Bnp$$

(60)

where r represents the nonequilibrium emission rate, r_0 the total equilibrium rate ($= r_{sp} + r_{st}$), and B the recombination constant given by

$$B = \frac{r_0}{n_i^2}$$

(61)

As the system approaches thermal equilibrium, $np \rightarrow n_i^2$, and r then approaches the equilibrium value. If we define Δn and Δp as the concentration of excess carriers and n_0 and p_0 as the concentrations at equilibrium, we can write

$$\Delta n = n - n_0 \qquad \Delta p = p - p_0$$

(62)

At equilibrium $r_0 = g_0$, and if we assume $g = g_0$ the increase in recombination rate is then

$$r - g = r_0 \left(\frac{np - n_i^2}{n_i^2} \right) = \frac{p_0 \, \Delta n + n_0 \, \Delta p + \Delta n \, \Delta p}{n_i^2} r_0$$

(63)

With this equation, we now look at the two methods for creating excess carriers: optical absorption and injection through a p-n junction.

Excess Carriers Created Through Optical Excitation

For optical band-to-band absorption, an equal number of holes are created as electrons, i.e., $\Delta n = \Delta p$ and we can write the recombination lifetime as

$$\tau_r = \frac{\Delta n}{r - g} = \frac{n_i^2}{r_0(n_0 + p_0 + \Delta n)} \tag{64}$$

$$= \frac{1}{B(n_0 + p_0 + \Delta n)} \tag{65}$$

For a p-type semiconductor, $p_0 \gg n_0, \Delta n$:

$$\tau_r = \frac{1}{Bp_0} = \frac{1}{BN_A} \tag{66}$$

Likewise, for n-type material

$$\tau_r = \frac{1}{BN_D} \tag{67}$$

In diodes fabricated in GaP doped with nitrogen, an increase in doping decreases radiative lifetimes up to a doping level of $2 \times 10^{18} \text{ cm}^{-3}$, above which nonradiative lifetimes begin to decrease due to an increase in Auger and deep-level nonradiative recombination [2].

For intrinsic material $n_0 = p_0 = n_i$. The lifetime is then a maximum for $\Delta n \ll n_0, p_0$ and a is minimum for $\Delta n = n_i$:

$$\tau_{r(\max)} = \frac{1}{2Bn_i}$$

$$\tag{68}$$

$$\tau_{r(\min)} = \frac{1}{3Bn_i}$$

Under steady-state illumination, $dn/dt = 0$ and τ_r can be defined as the time constant of excess carrier decay when the illumination is turned off. The time constant varies with the inverse of doping.

Excess Carriers from Carrier Injection

Under injection, electrons will be minority carriers in the p-type side, and the holes minorities in the n-type side. For example, into the p-region $p_0 \gg \Delta n, n_0, \Delta p$ and Eq. 63 becomes

$$r - g = \frac{\Delta n(p_0)}{n_i^2} r_0 \tag{69}$$

and, determining the lifetime,

$$\tau_r = \frac{1}{Bp_0} = \frac{1}{BN_A} \tag{70}$$

and the result is the same as the radiative lifetime derived for excess carriers created through optical excitation. The recombination lifetime with no recombination centers therefore, depends on the doping level and the recombination coefficient. The coefficient depends upon the physics of the material.

Under high injection we have $\Delta n \approx p_0$, and the lifetime is simply

$$\tau = \frac{1}{B \, \Delta n} \tag{71}$$

which is dependent on the injected current level.

3.2 Recombination Through Centers

As mentioned earlier, the result of energy released through electron-hole recombination via a defect center may be photon emission, a series of phonons, or emission of an Auger electron. The only major difference between recombination through a center and band-to-band recombination is the affect of the center on the carrier lifetime.

Whether a center N_t is a trapping center or a recombination center depends on the relative size of the electron and hole *capture cross sections* for that center. The electron capture cross section σ_n is related to the probability that a free electron can be captured at that center. In fact, the cross section is defined through the capture rate. The rate of capture at a center is proportional to the free-electron concentration, n, and the probability that the center is not occupied by an electron. The proportionality constant is σ_n:

$$\text{capture rate} = nN_t(1 - f_t)\sigma_n\langle v_{th}\rangle \tag{72}$$

Now, N_t represents the concentration of the center, $\langle v_{th}\rangle$ the average thermal velocity of the carrier, and f_t represents the Fermi function evaluated at the trap level E_t:

$$f_t = \frac{1}{1 + \exp[(E_t - E_f)/kT]} \tag{73}$$

$N_t(1 - f_t)$ is therefore the probability that the recombination center is not occupied by an electron (and therefore occupied by a hole). The constant $C_n = \sigma_n\langle v_{th}\rangle$ is known as the *capture constant*. The capture cross section has units of area, and thus C_n represents the volume of carriers removed per unit time.

The capture cross section for electrons σ_n is much larger for positive recombination centers than for negative centers. Values usually range between 10^{-11} cm^2 for a Coulomb attractive process to 10^{-22} cm^2 for a Coulomb repulsive capture process. For capture processes with very small cross sections, electron capture is only significant at very low temperatures, where the re-emission rate of the carrier is lowered.

The rate of emission of the carrier is proportional to the probability the center is occupied by an electron:

$$\text{emission rate} = \nu N_t (f_t) e^{-(E_c - E_t)/kT} \tag{74}$$

where ν is a constant called the *attempt to escape frequency*, and the factor $\exp\{-(E_c - E_t)/kT\}$ represents the probability that the electron has enough energy to escape from the trap. An *emission rate constant e_n* can then be defined as

$$e_n = \nu e^{-(E_c - E_t)/kT} \tag{75}$$

We can determine e_n using the principle of detailed balance: at equilibrium, emission and capture processes must be equal. Equating Eqs. 72 and 74 and using

$$n = N_c e^{-(E_c - E_f)/kT} \tag{76}$$

for the electron density give us a relationship for the attempt frequency:

$$\nu = N_c \sigma_n \langle v_{\text{th}} \rangle = N_c C_n \tag{77}$$

So far, we have only discussed the capture and re-emission of an electron at a trapping center. We now want to concentrate on the electron capture and subsequent hole capture at a recombination center. We can model the rate of recombination through such a center using the Shockely-Read-Hall recombination theory. We first make some assumptions: (1) The center is originally neutral (X_t^0) and first captures an electron. (2) Following the electron, a hole is captured at the now-negative center; i.e.,

$$e^- + X_t^0 \rightarrow X_t^- \tag{78a}$$

$$X_t^- + h^+ \rightarrow X_t^0 \tag{78b}$$

Again, either or both capture processes may be radiative. The successful recombination at center X_t^0 requires symmetric capture cross sections σ_n and σ_p. We do not need to distinguish among positive, negative, or neutral trapping centers. All that identifies the recombination center is the relative size of σ_n to σ_p. Capture at a neutral center will be weak if there is a competing process where the electron can be trapped at a second, positive center that has a larger cross section.

The above equations for recombination must compete with the probability of re-emission from the center:

$$X_t^- \rightarrow X_t^0 + e^- \tag{79a}$$

and for the hole:

$$X_t^0 \rightarrow X_t^- + h^+ \tag{79b}$$

Since we have excess carriers, represented by Δn and/or Δp, the concept of a Fermi level must be somewhat altered. Earlier we used the Fermi level to describe both n and p in equilibrium, since they are related by $pn = n_i^2$. Since we are no longer in thermal equilibrium, we will use separate demarcations to describe n and p. We can redefine the new carrier densities as

$$n = N_c e^{-(E_c - E_{fn})/kT} \tag{80}$$

and

$$p = N_v e^{-(E_v - E_{fp})/kT} \tag{81}$$

where E_{fn} and E_{fp} are the *quasi*-Fermi levels. Note that these levels represent the *electrochemical potentials*, that is, the chemical potentials of the electrons and holes, respectively.

If n_t represents the occupancy of the trap ($= N_t f_t$) then we can write the change in occupancy as the capture rate minus the emission rate:

$$\frac{dn_t}{dt} = n C_n N_t (1 - f_t) - n_t C_n N_c e^{-(E_c - E_t)/kT} \tag{82}$$

The net rate of hole capture is

$$\frac{d(N_t - n_t)}{dt} = n_t p C_p - C_p N_t (1 - f_t) N_v e^{-(E_t - E_v)/kT} \tag{83}$$

In steady state:

$$\frac{dn_t}{dt} = \frac{d(N_t - n_t)}{dt} \tag{84}$$

We now introduce n_1 and p_1 such that $n = n_1$ when $E_{Fn} = E_t$ and $p = p_1$ when $E_{Fp} = E_t$. Equating Eqs. 82 and 83 and solving give us a relation for f_t:

$$f_t = \frac{n C_n + p_1 C_p}{C_n(n + n_1) + C_p(p + p_1)} \tag{85}$$

Normally, recombination is dominated by one center. Then we can equate the recombination rate r to the capture rate in the center:

$$r = \frac{n(1 - f_t)}{\tau_{n0}} \tag{86}$$

We can also equate the generation rate to the electron emission rate from the center:

$$g = \frac{n_1 f_t}{\tau_{n0}} \tag{87}$$

where

$$\tau_{n0} = \frac{1}{\sigma_n \langle v_{th} \rangle N_t} \tag{88}$$

and a similar constant can be defined for holes:

$$\tau_{p0} = \frac{1}{\sigma_p \langle v_{th} \rangle N_t} \tag{89}$$

The constant τ_{n0} represents the lifetime of an electron when N_t is completely occupied by holes, and τ_{p0} represents the lifetime of a hole when N_t is totally occupied by electrons. The recombination rate is then solved as

$$r - g = \frac{pn - n_i^2}{\tau_{p0}(n + n_1) + \tau_{n0}(p + p_1)} \tag{90}$$

The recombination rate is proportional to the difference between the product pn and n_i^2 and rightly goes to zero as the pn product approaches n_i^2.

We can substitute $\Delta p = p - p_0$ and $\Delta n = n - n_0$:

$$r - g = \frac{n_0 \Delta p + p_0 \Delta n_0}{\tau_{p0}(n + n_1) + \tau_{n0}(p + p_1)} \tag{91}$$

3.3 Recombination away from the Depletion Region

We can now compute the recombination rate for our p-n junction containing recombination centers. In Section 2 in the n-type side away from the depletion region the equilibrium electron (majority) carrier concentration was defined as n_n and the hole (minority) concentration as p_n. Under low injection, there are no excess carriers in the conduction band ($\Delta n = 0$) and $n_n \gg p_n$. If we assume the recombination center is far from the conduction band, then $n_n \gg n_1$. Under low injection conditions we have

$$n \gg \Delta p \tag{92}$$

and the net recombination rate is simply

$$r - g = \frac{\Delta p}{\tau_{p0}} \tag{93}$$

Since the minority carrier lifetime is defined as

$$\tau_p = \frac{\Delta p}{r - g} \tag{94}$$

we have

$$\tau_p = \tau_{p0} \tag{95}$$

for these conditions. This makes sense, since the Fermi level (or more correctly the quasi-Fermi level) is far above the trap level, the center is occupied with electrons for the majority of the time, and the recombination rate is limited only by the lifetime of the injected holes. If, however, the recombination center is close to the conduction band, $n_n \cong n_1$, and

$$\tau_p \cong 2\tau_{p0} \tag{96}$$

and the recombination rate is significantly slower. This still requires a high electron occupation probability.

Under conditions of high injection, we have

$$\Delta p \cong n_n \tag{97}$$

and

$$\Delta p \gg n_i \tag{98}$$

If the level is near midgap

$$\Delta p \gg p_1, n_1 \tag{99}$$

then the recombination rate can be expressed as

$$r - g = \frac{\Delta p}{\tau_{p0} + \tau_{n0}} \tag{100}$$

and the lifetime τ_p becomes

$$\tau_p = \tau_{p0} + \tau_{n0} \tag{101}$$

Similar relations can be made for recombination through a single level on the p-side. Thus, we see minority carrier lifetime is determined by the relative position of the quasi-Fermi level and the trap level, as well as the position of the quasi-E_F with respect to the band edge. The lifetime is at a minimum under low injection if the center is at midgap. The lifetime is maximum under high injection and/or if the center is close to a band edge.

3.4 Recombination-Generation within the Depletion Region

Equation 91 was derived without regard for recombination-generation within the depletion region. This is an important omission, particularly for compound semiconductor devices where depletion-region recombination may be a significant component of the recombination current. Rewriting Eq. 26 for the concentration of electrons on the edge of the depletion region at the p-side gives

$$n(0) = n_p e^{qV_{app}/kT} \tag{102}$$

Multiplying both sides by p_p and remembering that $p_p n_p = n_i^2$, we arrive at an np product in the depletion region:

$$(np)_d = n_i^2 e^{qV_{app}/kT} \tag{103}$$

where we have assumed the majority carrier concentration is constant through the depletion region.

Substituting into Eq. 91 we obtain at the depletion edge that

$$(r - g)_d = \frac{n_i^2(e^{qV_{app}/kT} - 1)}{\tau_{p0}(n + n_1) + \tau_{n0}(p + p_1)} \tag{104}$$

Recombination increases exponentially with applied forward bias (up to high injection). In reverse bias, $V_{app} \ll 0$ and since there are no free carriers, $r - g$ can be written simply as the thermal generation of carriers:

$$g_d = \frac{n_i^2}{\tau_{p0}n_1 + \tau_{n0}p_1} \tag{105}$$

It is this generation component which is responsible for the reverse saturation current J_0 when a single-level recombination center is present. This again gives us the Shockley equation:

$$J = J_0(\exp(qV_{app}/kT) - 1) \tag{106}$$

only with a slightly larger value of J_0 given by $J_0 = qWg_d$, where W is the depletion width.

3.5 Surface Recombination

The importance of surface or interfacial recombination can never be understated when addressing LED operation since it is the dominant nonradiative path and thus controls the nonradiative carrier lifetime. Surface recombination in III-V compounds seems simple at first glance: nonradiative recombination through surface rather than bulk recombination states in the band gap. Thus, surface recombination can be described as an extension of bulk theory. As was demonstrated for the bulk lifetime in Section 3.3, the surface recombination lifetime will depend on the relative positions of E_F and E_t in the gap. For a single center, the effect of the energy quantities $E_c - E_F$ and $E_t - E_F$ on the surface lifetimes is similar to that of the bulk.

At the surface, a value of $E_c - E_F$ different from that in the bulk semiconductor means there is band bending at that surface. It is well known that a surface barrier exists in compound semiconductors as well as a large quantity of surface states and both determine the value of the nonradiative lifetime. The band bending is such that when recombination exists at the surface, a hole current J_p^s flows toward the surface and an electron current J_n^s flows away. As is customary with all extraneous currents, if a reliable model is to be developed, these surface currents must be added to the total current before solving the continuity equation. Shockley [3] defined the surface recombination velocity for electrons, S_n, and for holes, S_p, as

$$J_n = -q \, \Delta n S_n \tag{107a}$$

and

$$J_p = q \, \Delta p S_p \tag{107b}$$

For a heterostructure device such as the LED discussed in Section 5, the recombination is dependent upon interfacial states between epitaxial layers. For epitaxy, each epitaxial layer will have a different lattice parameter a, and a *misfit strain* is produced due to the mismatch (Δa) of the layers. The relation between misfit strain and *interfacial* recombination velocity is found experimentally to be [4]

$$S \cong (2 \times 10^7) \frac{\Delta a}{a} \text{ cm}^2/\text{s} \tag{108}$$

4 LED EFFICIENCIES

The basic properties which characterize LEDs include the following: (1) spectral output, (2) output power, (3) rise and fall times, (4) angular emission pattern. It is common to define the radiant output power efficiency η_p as

$$\eta_p = \frac{\text{power of radiation emitted}}{\text{electrical power consumed}} \tag{109}$$

This efficiency is easily measured by integrating the power output around the solid angle 4π. An *external quantum efficiency* can be defined as

$$\eta_{\text{ext}} = \frac{\text{number of photons emitted}}{\text{number of carriers passing the junction}} \tag{110}$$

If we multiply the denominator of Eq. 110 by qV_{app}, where V_{app} is the total applied voltage, the result is the electrical power consumed. In fact, we can relate η_{ext} to η_p:

$$\eta_p = \eta_{\text{ext}} \frac{h\nu}{qV_{\text{app}}} \tag{111}$$

where $h\nu$ is the emitted photon energy. For a heavily doped junction, values of the series resistance of neutral n and p regions are small, and, since $qV_{\text{app}} \approx h\nu$, $\eta_p \approx \eta_{\text{ext}}$.

There are several contributions to the external quantum efficiency. An *optical efficiency* η_{opt} can be defined as the fraction of photons generated that actually leave the crystal. An *injection efficiency* η_I is the injected current divided by the total current. The *internal quantum efficiency* η_{int} is the fraction of carriers injected into the neutral region that actually recombine with a resulting emission of photons. The external quantum efficiency can then be written as

$$\eta_{\text{ext}} = \eta_{\text{int}} \eta_{\text{opt}} \eta_I \tag{112}$$

In this section we will examine each of these components separately.

4.1 Injection Efficiency

As noted, there are several ways to define LED efficiencies. An *injection* efficiency can be defined for either holes or electrons as the ratio of minority carrier currents to total currents:

$$\eta_I = \frac{J_n}{J_p + J_n + J_{\text{sc}} + J_s} \tag{113}$$

where we have used minority carrier (electron) injection into the p-side as our example. J_p is the hole injection current into the n-side, while J_n is its counterpart for electrons in the p-side. These radiative and non-radiative components of the current in the neutral region, $J_p + J_n$, the depletion-region recombination current J_{sc}, and the surface recombination current J_s all add to the total current. At very high current densities, the diffusion current becomes much greater than J_{sc} or J_s, and injection efficiency increases with drive current. For an operable device we may ignore the J_{sc} and J_s components and substitute Equation 38 for the injection currents:

$$J_n = \frac{D_n n_p}{L_n} \exp\left(\frac{qV_{app}}{kT}\right) \tag{114}$$

and with the equivalent relation for J_p, the injection efficiency becomes

$$\eta_I = \frac{D_n n_p / L_n}{D_p p_n / L_p + D_n n_p / L_n} \tag{115}$$

$$= \left(1 + \frac{D_p L_n p_n}{D_n L_p n_p}\right)^{-1} \tag{116}$$

$$= \left(1 + \frac{\mu_p L_n p_p}{\mu_n L_p n_n}\right)^{-1} \tag{117}$$

where μ_n and μ_p are the electron and hole mobilities, respectively. We have again used the Einstein relation ($D/\mu = kT/q$) and $np = n_i^2$. Since $n_n \approx N_D$ and $p_p \approx N_A$, the injection efficiency thus depends on the relative doping of the junction. For GaAs the hole mobility is roughly 20 times smaller than that of the electron, and the electron injection efficiency is greater than that of the hole for the same level of doping. Thus, most light is generated in the p-side. The injection efficiency for holes can be made higher by preparing a p^+n junction, which also reduces series resistance of the p-region. However, in the heaviest-doped material, radiative recombination is actually lost due to an increase of nonradiative recombination pathways at high doping.

In heterostructure LEDs, only one carrier (electron or hole, depending on the structure of the device) is normally injected, and the injection efficiency can be made close to unity.

4.2 Internal Quantum Efficiencies

In Eq. 113 we have neglected to separate the nonradiative and radiative recombination paths in the equation for total current. Recombination that leads to no photon generation in the neutral n- and/or p-regions immediately after injection is included in the internal quantum efficiency.

We can define a radiative, or internal, quantum efficiency as

$$\eta_{int} = \frac{r_r}{r_r + r_{nr}} \tag{118}$$

where r_r and r_{nr} are the radiative and nonradiative recombination rates, respectively. By using the substitution

$$r = \frac{\Delta n}{\tau} \tag{119}$$

we can define η_{int} in terms of the lifetimes:

$$\eta_{int} = \frac{\tau_{nr}}{\tau_{nr} + \tau_r} \tag{120}$$

The radiative efficiency can be increased either by reducing the radiative or by increasing the nonradiative carrier lifetimes. The radiative lifetime for GaAs is small, on the order of 10^{-9} s for material doped to 10^{17} cm^{-3} [5].

For simple GaAs p-n junction LEDs, the major contribution to nonradiative recombination is where the depletion region intersects the surface, so the periphery of the light-emitting device is made as small as possible. For recombination in heterojunction LEDs, there is also a nonradiative component of the recombination current at the heterointerface of the junction to consider.

4.3 Optical Efficiency

Once photons are generated within the LED structure, they must somehow escape the device and reach the outside world. There are two major loss mechanisms which must be overcome for the photons to escape: (1) losses due to reabsorption of the emitted light, and (2) losses due to reflection, both direct backreflection (Fresnel loss) and total internal reflection.

In reabsorption, light is first emitted in an LED from a vertical transition in k-space. This transition may involve a recombination center or be a band-to-band transition across a direct band gap:

$$e^- + h^+ \rightarrow \text{null} + \Delta E \tag{121}$$

This equation represents the recombination of an electron and hole with the subsequent release of energy $\Delta E = h\nu$. Any downward transition may also be reversed: the transition may absorb a photon, and the electron-hole pair is regenerated. Energy loss mechanisms during optical emission dictate that an energy larger than ΔE is needed to reverse the process. However, this disregards the band tailing discussed in Section 11.1 and Stark effect of Section 11.2. The band-gap energy is reduced in areas where inhomogeneities exist and where electric fields are highest. The absorption edge is not sharp in these regions, and there is the possibility that a photon of energy ΔE will be reabsorbed. The extent of reabsorption is determined by the value of the absorption coefficient at the emission wavelength:

$$I = I_0 e^{-\alpha x} \tag{122}$$

For a direct-gap semiconductor, the emission wavelength is close to the absorption edge, and α in this region is high, on the order of 10^4 cm^{-1}. For every micron traveled, two-thirds of the emitted light is reabsorbed. For a GaAs homojunction LED the optical efficiencies, which are defined by the percentage of recombination-produced light that eventually escapes, are then seriously perturbed by reabsorption.

For GaP:N, the band gap is indirect and only the N center itself is free to absorb energy. Consequently, the absorption coefficient is lower than 10^2 cm^{-1} at the emission wavelength, and reabsorption is less thanks to the indirect band gap.

In some sense, absorption losses can be reversed, meaning the electron-hole pair can recombine again and create another photon. However, the efficiency of this process is unclear. Other mechanisms of absorption are also possible, including free carrier and impurity absorption. While band-to-band absorption is dominant in direct-gap LEDs, for GaP:N the major mechanism is absorption by the nitrogen impurity. Nitrogen in GaP absorbs up to 50% of the photons generated in these diodes [6].

In conventional diodes, losses to absorption are reduced by locating the *p-n* junction as close to the surface as possible. For optical fiber applications, a small spot less than or equal to the fiber diameter ($\sim$50 μm) is needed to launch the maximum amount of power into the fiber. This has been done by growing very thin *p*-layers and by forming a small-diameter *p*-contact to promote crowding of the current in a very small area. A well is then etched in the substrate and the fiber is attached as close to the junction as possible. In some cases, the substrate is etched to remove the bulk of its thickness.

Absorption losses can be reduced by shifting emission energies to values below the band gap. This can be done by two methods: doping and double heterostructure.

Losses due to reflection include direct backreflection (Fresnel loss) and total internal reflection. Of the two, the latter is far more serious for a semiconductor/air interface. All radiation is reflected backward for light impinging the interface at a critical angle greater than θ_c given by

$$\theta_c = \sin^{-1}\left(\frac{n_1}{n_2}\right) \tag{123}$$

where θ_c is measured normal to the surface, n_2 represents the index of refraction of the semiconductor, which is $\sim$3.3 at the emission wavelength, and n_1 is the index of air. For GaAs, this corresponds to a narrow cone of radiation of $\theta_c \approx 16°$. The angle can be increased by encasing the device in an epoxy hemisphere of higher n_1.

For an edge emitter, reflection losses are also great, but the exact emission cone is determined by the width of the active region.

Transmission losses from reflection are most easily reduced in display LEDs by shaping the emitting surface into a sphere or hemisphere. This lowers the probability of incidence at an angle greater than the critical angle for total internal reflection and reduces such reflections. The diode can also be encapsulated in a high-index plastic and coated with antireflection coatings.

5 MATERIALS FOR LEDs

5.1 GaP and GaAsP

The semiconducting alloys GaP and $GaAs_yP_{1-y}$ have been workhorses for the commercial visible LED industry for many years. Few people are not familiar with visible red, yellow, or green LEDs fabricated from these materials. Low-cost fabrication and device processing of GaP and $GaAs_yP_{1-y}$ *p-n* junctions on GaAs and GaP substrates have provided the world with a wide range of commercial visible displays. Low-cost GaP and $GaAs_yP_{1-y}$ diodes have reasonable injection, luminescent, and external quantum efficiencies for the simple visible structures they represent.

For band-to-band radiative recombination, $h\nu \approx E_g$ and LED materials for visible spectrum applications (0.4 to 0.7 μm) must have band-gap energies ranging from ≈ 1.8 to 2.8 eV. For radiative emission through a trapping center (level-to-band, band-to-level, or level-to-level) a direct band gap is unnecessary, but the radiation energy, $h\nu$, will be smaller than E_g. Since GaP has an indirect band gap of 2.26 eV at 300 K, radiative recombination centers must be introduced to obtain optical emission from GaP LEDs. Such centers are provided by doping with nitrogen or zinc + oxygen. Green GaP LEDs of emission wavelength $\lambda = 0.565$ μm are fabricated by doping both sides of a GaP *p-n* junction with nitrogen. Because nitrogen resides on the phosphorus sublattice, it forms an uncharged recombination center within the GaP band gap. Uncharged impurities are called *isoelectronic*. An increase in nitrogen content shortens the radiative lifetime and thus luminescent efficiencies are increased as explained through Eq. 120. Radiative efficiencies through the center are still rather low as compared to LEDs that support band-to-band recombination, but extremely high efficiencies are unnecessary because of the high sensitivity of the human eye to green wavelengths. Still, much care is taken in device design to minimize nonradiative recombination paths. Red GaP LEDs are also available. Doping GaP with zinc and oxygen produces donor oxygen and acceptor zinc energy levels, from which red GaP recombination is produced.

The ternary $GaAs_yP_{1-y}$ is a direct-gap semiconductor for alloy compositions $y \leq 0.45$, and for these compositions band-to-band radiative recombination (red) diodes can be prepared. Yellow and orange LEDs fabricated from $GaAs_yP_{1-y}$ must be produced by isoelectronic doping. Radiative efficiencies of band-to-band $GaAs_yP_{1-y}$ red diodes are low compared to GaP:N diodes due to a higher concentration of nonradiative recombination centers in the ternary, compared to the binary structure. External efficiencies are also lower for the ternary due to higher reabsorption in the direct-gap material. (See Section 4.3.)

Figure 5 shows a surface-emitting GaAsP/GaP structure for commercial use. The diffused p-region is as narrow as possible to minimize absorption losses. In addition, the surface can be made hemispherical to minimize reflection losses.

5.2 Materials for Heterojunction LEDs

Creation of any heterojunction device means the creation of an interface which must always be considered in any treatment of the device physics. We are most concerned with minimizing any nonradiative paths introduced by the interface. Interfacial recombination is treated as a form of

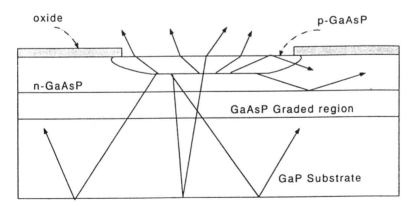

Figure 5   Surface-emitting GaP/GaAsP LED. (From Ref. 7.) A GaAsP graded region is prepared to minimize strain in the emitting region. The GaP substrate is "transparent" to the GaAsP-emitted light and helps to reflect light out of the device.

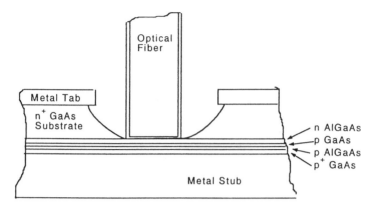

Figure 6   Burrus-type LED as used for a fiber-optic source. (From Ref. 8.)

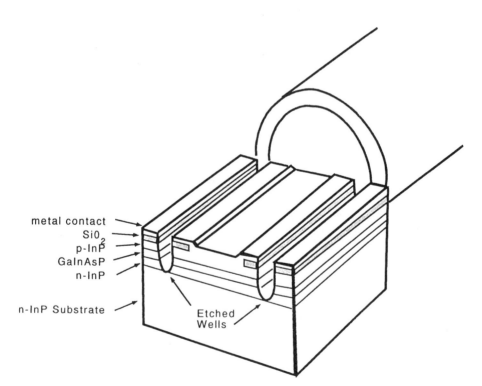

Figure 7   Edge-emitting LED as would be used for a fiber-optic source. (From Ref. 9.)

surface recombination, as explained in Section 3.5. For heterostructure devices, a close lattice match between the two semiconductors that make up the interface is necessary to reduce the concentration of any misfit defects and to minimize strains in the interface as the temperature varies. If any high-temperature processing is involved in preparation of the device, the thermal expansion coefficient for that temperature range must also be matched.

The ternary $Al_xGa_{1-x}As$ is useful for heterostructure devices as it can be lattice-matched to GaAs substrates for almost all compositions. It is only direct for $x < 0.45$, it does not have an isoelectronic radiative center, and thus it is limited to use for the 800–900 nm optical spectrum. Figure 6 is of a "Burrus-type" LED common for fiber-optic communications. Layers of epitaxy provide the emitting junction. The contact is a reflecting surface to reflect any light back toward the optical fiber. A hole the proper depth is etched in the substrate for insertion of the optical fiber.

AlGaAs will not provide light for an essential range of wavelengths: 1.35–1.65 micron. For this range a narrower band gap of the quaternary GaInAsP is needed. Figure 7 shows an edge emitter configuration for fiber-optic applications, often used for coupling light from GaInAsP LEDs.

6 HIGH-RADIANCE HETEROSTRUCTURE LEDs

Many of the severe limitations in external quantum efficiencies that plague homojunction LEDs can be reduced in heterojunction LEDs. The double-heterostructure (DH) devices shown in Figs. 6 and 7 do not suffer from the same amount of self-absorption as homojunction devices. This comes at the expense of more complex processing and a slightly lower internal quantum efficiency. Furthermore, these devices provide high optical power without the complexities of operation and fabrication of laser diodes.

The DH device operates chiefly on the concepts of *confinement*. Confinement of carriers and confinement of photons are both realized. The first is a outcome of the discontinuities of band gap; the second is a result of the waveguide action of the device. In a double-heterostructure device, a thin (2–2.5 μm) expitaxial *active region* is encased on each side by *cladding regions*. The physical device is shown in Figs. 6 and 7. The cladding regions are made of semiconductors that have band-gap energies larger than the active region. We can draw "ideal" *p-n* junction energy bands of a double-heterojunction device in Fig. 8, which is a modified drawing of Fig. 1. The modification in the bandstructure comes from a "kink" in the energy bands where the active region is surrounded by the wider band gap of the cladding. For an ideal DH diode, after injection the

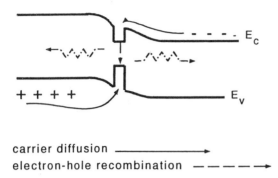

carrier diffusion ──────────▶
electron-hole recombination ─ ─ ─ ─ ─▶
photon emission _·/ˈ\ˑ·/ˈ_▶

Figure 8 "Idealized" double-heterostructure *p-n* junction. Injected carriers are trapped in an energy "well," where radiative recombination is primarily confined.

electrons and holes do not diffuse apart as they do in the homojuction, but are trapped in the kink in the energy bands.

Carrier confinement results in high carrier densities which reduce the radiative lifetimes and thus improve frequency response. Furthermore, since the energy of the emitted photon is less than the band-gap energy of the cladding regions, losses due to absorption in the cladding regions are minimized.

The actual interface between active region and cladding is difficult to model and depends on the properties of the respective materials and the ideality of the interface. A model that is often used to illustrate the interfacial properties is the Anderson model. Figure 9 shows a DH *n*-AlGaAs/*p*-GaAs/*p*-AlGaAs energy band structure based on such a model [10]. The energy barrier ΔE_c will act as a confinement for electrons in the gallium arsenide and the barrier ΔE_v for holes.

The transition from DH LEDs to DH laser diodes is relatively minor. The major difference in the devices is the exactness of the fabrication process. Because of the similarities between DH laser diodes and LEDs, what is covered in Section 6 holds for both devices.

6.1 Frequency Response of Double-Heterostructure LED

We can relate the modulation bandwidth of a DH LED to four major device parameters that have already been defined:

Width of the active region, W
Minority carrier lifetime, τ_n and τ_p

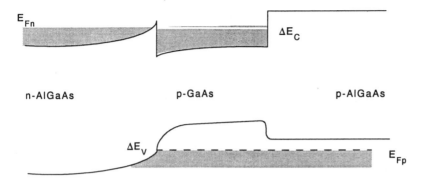

Figure 9 Anderson model of an AlGaAs/GaAs/AlGaAs double-heterostructure device.

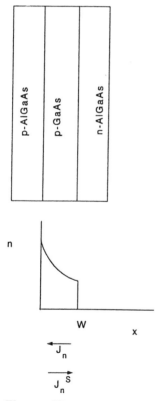

Figure 10 Spatial dependence of minority carrier concentration in a double heterojunction, with surface (interface) recombination included. (From Ref. 11.)

Reflection losses and reabsorption (external efficiencies)
Relative doping of the *p-n* junction

Because of the presence of the two interfaces, we must first realize that a major component of the current will be surface recombination. We first recall Eq. 107a, which defined the surface recombination velocity:

$$J_n^s = -q[\Delta n]S_n \tag{124}$$

The spatial dependence of the minority carrier concentration will appear as shown in Fig. 10. Here we are considering a *p*-type active region. Because the width W is quite small and the injected current large, we can safely make the assumption

$$\Delta n \approx n \tag{125}$$

At the *p-n* junction inteface we subtract Eq. 124 from the injected electron current:

$$J_n = -qD_n \frac{dn}{dx} + qS_n n \tag{126}$$

Evaluating this equation at the two interfaces gives the following boundary conditions:

$$\left.\frac{dn}{dx}\right|_{x=0} = \frac{J_n}{qD_n} - \frac{S_n}{D_n} n(0) \tag{127}$$

and

$$\left.\frac{dn}{dx}\right|_{x=w} = -\frac{S_n}{D_n} n(W) \tag{128}$$

where we have assumed the surface recombination velocities to be equal at both interfaces. The spatial dependence of the electron density can be found from evaluating the continuity equation in steady state [11]:

$$n(x) = \frac{JL_n}{qD_n}\left\{\frac{\cosh\left(\frac{W-x}{L_n}\right) + (L_nS_n/D_n)\sinh[(W-x)/L_n]}{[(L_nS_n/D_n)^2 + 1]\sinh(W/L_n) + (2L_nS_n/D_n)\cosh(W/L_n)}\right\} \tag{129}$$

where L_n is the electron diffusion length. The average electron density in the active region is

$$\langle n \rangle = \frac{1}{W}\int_0^W n(x)\,dx = \frac{J\tau_{DH}}{qW} \tag{130}$$

where a new lifetime τ_{DH} is defined from

$$\tau_{DH} = \tau_n \left\{ \frac{\sinh(W/L_n) + (L_n S_n/D_n)[\cosh(W/L_n) - 1]}{[(L_n S_n/D_n)^2 + 1]\sinh(W/L_n) + (2L_n S_n/D_n)\cosh(W/L_n)} \right\}$$

(131)

$$\tau_n = \frac{L_n^2}{D_n}$$

The quantity τ_n is the normal carrier lifetime expected without any interface present, and τ_{DH} is now the average lifetime with interfacial recombination included. If the surface recombination velocity is less than the diffusion rate, $L_n S_n/D_n \ll 1$ and $W/L_n \ll 1$, and the lifetime reduces to

$$\frac{1}{\tau_{DH}} = \frac{1}{\tau_n} + \frac{2S}{W}$$

(132)

Since the materials are direct gap, most of the recombination without the interface present and in the absence of any recombination centers is radiative and $\tau_n \cong \tau_r$. For the device with an interface present we can write

$$\frac{1}{\tau_{DH}} = \frac{1}{\tau_r} + \frac{1}{\tau_{nr}}$$

(133)

and τ_{nr} now represents the nonradiative component of the lifetime due to the presence of the interface. From Eq. 120 we have [11]

$$\eta_{int} = \frac{\tau_{DH}}{\tau_r} = \left(1 + \frac{2S\tau_r}{W}\right)^{-1}$$

(134)

If $W = 0.3\ \mu m$ and $\tau_r \cong 10^{-9}$ s for $\eta_i = 50\%$ requires that $S \leq 2 \times 10^{-4}$ cm/s. According to Eq. 108, $\Delta a/a_0$ must then be less than 10^{-3} [12]. This requires composition control for AlGaAs across the wafer of better than 10% AlAs, and for GaInAsP control of better than 2% across the wafer of InP.

7 LED DEGRADATION

A favorable aspect of LEDs and laser diodes is their reliability. LEDs have lifetimes estimated at $\sim 10^6$–10^9 h. Most degradation effects originate from the relatively large (as compared to crystalline silicon or germanium) quantity of crystalline defects present in compound semiconductors. Both dislocations (line) and point defects (similar to those discussed in Chapter

1) are present. Dislocations propagate through epitaxial layers from the substrate during epitaxial growth, and the density of these defects tends to vary across the surface of the GaAs wafer. Devices fabricated close to the perifery of the wafer may be less reliable if a high concentration of dislocations is present in those areas. Many catastrophic failures in LEDs can be filtered out at fabrication time by completing a 100-h "burn-in" where diodes fabricated in high-defect areas of the wafer are immediately identified. Devices which pass the burn-in test suffer from much slower failure mechanisms which only affect operation on a long-term basis.

The movement of dislocation under forward bias is the main failure mechanism in devices based on the AlGaAs/GaAs and GaAsP alloy systems. Such dislocations are called "dark line defects," a name given from the signature of the electroluminescent image of the device; a signature which directly identifies the defect. Such dislocations are decorated with defects or impurities which act as nonradiative recombination centers, and in a very short time sharply reduce the output of the LED or laser diode where they reside.

Dark line defects can propagate through a multilayer structure by dislocation climb. The mechanism for climb is related to the nonradiative recombination process, and climb is not normally observed if the device is not undergoing recombination. The nonradiative recombination mechanism involves a defect center, where energy released through recombination is localized temporarily at that center [13]. The amount of energy is sufficient to cause defect motion, and as a result the defect starts to move away from the junction. The result in GaAs/AlGaAs and GaAsP diodes is a dislocation "climb," or movement of the dark line defect.

Since the energy for motion is derived from recombination, and that energy released is dependent on the size of the energy gap, materials with wide gaps have more energy available for recombination-enhanced failure. It has been shown in GaAsP and InGaAs that the wider gap of the alloy the higher the failure rate [14]. Wide-gap compounds also tend to have a larger concentration of defects at growth at which such failures can be initiated.

Dark line defects are reduced by using GaAs substrates with low dislocation densities and by reducing some of the strain induced by semiconductor processing. Many improvements have come from reducing processing-induced strain: reducing the number of high-temperature processing steps, replacing hard contacts with ductile material, reducing dicing-induced strain, and improving the lattice match of epitaxial layers.

Since the quaternary GalnAsP has a smaller band gap, the energy available through recombination is less and recombination enhanced defect motion is not observed in this system. Since the wide-gap material is the binary lnP and not ternary or quaternary material, the thermal conductivity of this active region is also higher, thus reducing defect activity.

REFERENCES

1. J. S. Blakemore, *Semiconductor Statistics*, Dover, New York, 1987, Chapter 5.
2. M. H. Pilkuhn, Light emitting diodes, in *Handbook on Semiconductors*, Vol. 4 (T. S. Moss and C. Hilsum, eds.), North-Holland, Amsterdam, 1981, p. 589.
3. W. Shockley, *Electrons and Holes in Semiconductors*, Van Nostrand, Princeton, 1950, p. 321.
4. H. Kressel, The application of heterojunction structures to optical devices, *J. Electron. Mater. 4*, 1081 (1975).
5. M. S. Tyagi, *Introduction to Semiconductor Materials and Devices*, Wiley, New York, 1991, p. 359.
6. R. Z. Bachrach, W. B. Joyce, and R. W. Dixon, Optical coupling efficiency of GaP:N green light emitting diodes, *J. Appl. Phys., 44*, 5458 (1973).
7. R. Kniss, GaP under GaAsP brightens LED colors, *Electronics*, May 2, 1974, p. 34.
8. C. A. Burrus, Electroluminescent diodes coupled to optical fibers, *Proc. IEEE, 59*, 1263 (1971).
9. R. C. Goodfellow, A. C. Carter, I. Griffith, and R. R. Bradley, GalnAsP/lnP fast, high radiance, 1.05–1.3 μm wavelength LEDs with efficient lens coupling to small numerical aperture silica optical fibers, *IEEE Trans Electron. Devices, ED-26*, 1215 (1979).
10. A. G. Milnes and D. L. Feucht, *Heterojunctions and Metal Semiconductor Junctions*, Academic Press, New York, 1972, Chapters 1 and 2.
11. T. P. Lee and A. G. Dentai, Power and modulation bandwidth of GaAs–AlGaAs high radiance LEDs for optical communications systems, *IEEE J. Quantum Electron., QE-14*, 150 (1978).
12. H. Kressel, M. Ettenberg, J. P. Wittke, and I. Ladany, Laser diodes and LEDs for fiber optical communication, in *Semiconductor Devices for Optical Communication, Topics in Applied Physics*, vol. 39 (H. Kressel, ed.), Springer-Verlag, Heidelberg, 1982, p. 11.
13. L. C. Kimmerling, Recombination enhanced defect reactions, *Solid State Electron., 21*, 1391 (1978).
14. M. Ettenberg and C. J. Neuse, Reduced degradation in lnGaAs electroluminescent diodes, *J. Appl. Phys., 46*, 2137 (1975).

3

Semiconductor Lasers

1 INTRODUCTION

It was only natural for laser diodes to evolve from light-emitting diode technology. The two devices are very similar, and the laser diode owes its existence to the original studies on LEDs. The laser diode market has, since inception, far surpassed that of any other laser source. With it has come several new technologies as well, including the diode-pumped solid-state laser, laser diode arrays, parallel lightwave communications, and compact disk technology.

The laser diode benefits from its small size, low cost, and high output power for such a small optical resonator. It obtains this advantage from the tremendous optical gain available per resonator length. Its only major drawbacks are the extreme divergence in the exiting light due to such a small resonator, and a lack of coherence in the light as compared to equivalent-powered gas lasers. The coherence problem is also a result of the small cavity length.

Diode lasers are less expensive, more compact, and more reliable than other laser systems. But perhaps their greatest advantage over other lasers is the fact they can be modulated simply by turning a current on and off. In principle, the switching time of lasers is only limited by the carrier lifetime, and thus high modulation rates are possible.

The main reason for replacing LED applications with laser diodes is the increase in speed and power density. The lifetime of free carriers (electrons and holes) under conditions of stimulated emission is much shorter than the spontaneous lifetime; thus LED applications are limited to ~100 MHz, while response times of laser diodes are less than 1 ns. When focused, laser diodes are capable of delivering several milliwatts of power into a very small spot. Such performance is needed for certain applications such as long-haul fiber optics and compact disk read/write operations.

Laser diodes enjoy advantages of portability, high power in a small size, ease of modulation, and low cost because devices can be fabricated with monolithic techniques. The possibilities of advances such as high-frequency modulation and optical wavelength tunability of devices are also promising. Laser diode technology continues to improve with ever-increasing higher output powers and lower divergence while maintaining optical mode stability and diffraction-limited conditions.

The physics of radiative recombination, both stimulated and sponta-neous, in a semiconductor have been discussed in Chapter 2, along with the basics of *p-n* junction physics. Because of the similarities of LEDs and laser diodes, the majority of equations derived in that chapter hold here, and many will be used again.

1.1 Laser Diode Basics

Laser diodes operate on the same premise of the high-brightness LED, with some notable differences. In Section 6 of Chapter 2 the concepts of "carrier confinement" were introduced. Using a double-heterostructure (DH) device such as that illustrated in Fig. 9 of Chapter 2, carriers are confined to an *active region* of the diode. Like the DH LED of Section 6 of Chapter 2 the concepts of confinement of carriers and confinement of the emitted electromagnetic energy is important for laser diode operation.

For an LED to lase, the same two basic conditions of other laser sys-tems must be met: (1) population inversion (gain) and (2) optical feedback. Confinement of carriers is necessary to achieve population inversion, and confinement of photons, or waveguide action, is necessary for the emitted light to be fed back through the gain region instead of being lost to the surrounding semiconductor. Photon confinement is provided by a high-refractive-index region near the region of population inversion.

Population inversion is achieved in the recombination region, the active region of the laser diode where optical gain is produced by stimulated electron-hole recombination. Figure 1 illustrates the active region, and how population inversion is achieved in a DH diode where a high injected current density is needed to create the condition. When injected current

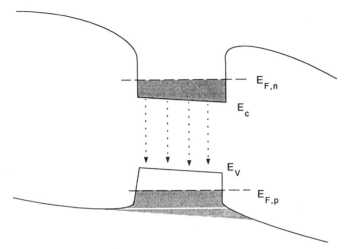

Figure 1   Idealized double heterostructure and relative positions of the quasi-Fermi levels ($E_{F,p}$, $E_{F,n}$), within the active region under population inversion conditions.

density is sufficient, the amount of stimulated emission produced is greater than the absorption. Laser oscillations occur when the round-trip gain in the active region exceeds the total losses experienced in the same distance. Principle loss mechanisms are scattering from inhomogeneities in the semiconductor, absorption, and mirror losses.

The condition for population inverstion in semiconductors is the primary difference of laser diodes from other laser systems. In other systems, electromagnetic energy is produced by transitions between discrete energy levels. In semiconductor lasers, electromagnetic energy is produced from radiative recombination between a continuum of energy levels.

The condition for population inversion in a semiconductor is shown in the energy band diagram of Fig. 1. The states in the conduction band are filled with electrons up to a level E_{Fn}, while those in the valence band are empty of electrons above a level E_{Fp}. In that figure, E_{Fn} represents the quasi-Fermi level for electrons, and E_{Fp} the quasi-Fermi level for holes. The incident light of intensity $I(\omega_0)$ induces a transition from a level E_a to level E_b. The resulting relaxation from E_b to E_a results in contributing one photon back to the beam. In a gas or conventional solid-state laser, E_b and E_a are descrete, the population inversion condition is governed by classical Boltzmann statistics, the population of states is small and dependent on the amount of material in the cavity. In a semiconductor the density of states is large and occupancy probabilities are governed by Fermi-Dirac statistics. The number of states is roughly equal to the number of atoms in the semiconductor. Thus, the transition probability from a to b is proportional to the product $f(E_a)[1 - f(E_b)]$ where $f(E)$ is the Fermi function given by Eq. 24 in Chapter 1. In the figure, E_{Fn} is *degenerate* because it resides in the conduction band, while E_{Fp} is degenerate in the valence band. Later, it will be shown that population inversion exists when

$$E_{Fn} - E_{Fp} > h\nu$$

Only frequencies whose photon energies are smaller than the quasi-Fermi level separation are amplified.

To obtain the conditions necessary for population inversion, a specialized structure must be fabricated to ensure confinement of carriers and photons. Referring to the device of Fig. 2, confinement is needed in the transverse and lateral dimensions of the laser, where the light should not exit. A feedback mechanism is necessary in the longitudinal direction, which is the direction in which the light should exit. Because so many different possibilities exist for diode design, only a few select will be mentioned here.

1.2 Basic Structures

Figure 3 shows various structures that are used to produce the carrier and photon confinement in the transverse dimensions. The refractive

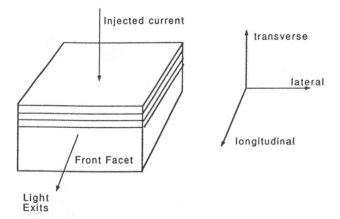

Figure 2 Relative directions in a laser diode. Light is emitted primarily from the front facet.

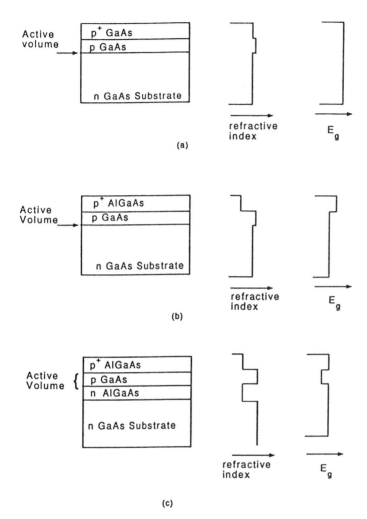

Figure 3 Various transverse configurations, including (a) homojunction, (b) single-heterojunction, and (c) double-heterojunction laser diodes.

index and band-gap profiles are also shown in Fig. 3. Photon confinement is usually realized by producing an internal waveguide in the active region. To confine the stimulated radiation produced to the region producing the gain requires a high-refractive-index region near the active region. Photons are then confined to the high-index regions by the lower index *cladding* regions.

The original *homojunction* design of Fig. 3a was simple and inexpensive to manufacture. It relied on high injection alone to achieve population inversion, and waveguide action was dependent on an increase in the refractive index with current density. The result was poor photon and carrier confinement and hence a high threshold current. Lasing could only be established in homojunction diodes by pulsing the injected current. In the single-heterostructure device of Fig. 3b, (also known as the close-confinement laser) the band-gap energy of the p^+ layer is higher than the active region. This provides a confinement barrier for electrons in the *p*-side and a roughly 5% decrease in refractive index in the AlGaAs. The device is only slightly more difficult to fabricate than the homojunction. However, at the *n-p* interface, there is only a small change in the refractive index. Thus, losses into the *n*-region are heavy, and an increase of current is necessary to increase the refractive index difference between *n* and *p* GaAs.

Homo- and single-heterostructure lasers have been completely supplanted by the double-heterostructure device. The increase in difficulty

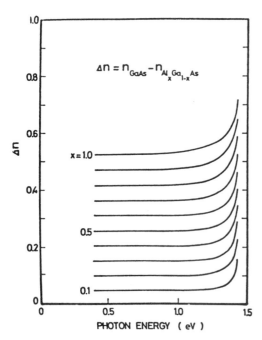

Figure 4   Refractive-index step variation of $Al_xGa_{1-x}As$ as a function of the photon energy with x increments of 0.1 [1].

of manufacturing the DH device is outweighed by the benefits of having two heterointerfaces. The DH device of Section 6 in Chapter 2, combined with a feedback mechanism, creates the DH laser diode. The "ideal" double heterostructure of Fig. 3c provides both carrier and photon confinement. Carrier confinement is provided by the two heterointerfaces as described earlier. Photon confinement is provided by "waveguide" action, as described below.

Figure 4 [1] shows how, for AlGaAs, the refractive index decreases as the AlAs content increases in the alloy. The energy band gap meanwhile increases with AlAs content. These are convenient properties of the alloy, as the result is two AlGaAs cladding or confinement layers surrounding the active region. The active region is of GaAs, or AlGaAs of low AlAs content, surrounded by AlGaAs of higher AlAs content in the cladding. Photon confinement is provided by an increase in refractive index as the photons travel from the smaller to the wider-band-gap material. The result is confinement to the high-gain active region.

2 LASING CONDITIONS

When designing the lateral and transverse dimensions of the laser, a trade-off must be met between the width of the lasing spot, the threshold current, and the output power. As either the width in the lateral, or the thickness in the transverse dimension decreases, there is an increase in the width of the lasing spot in that dimension due to the increased amount of light that travels in the cladding. This is accompanied by a higher possible output power, as small spot size means output power must be reduced to avoid facet damage. Wider spot size also means lower output beam divergence. In addition, for relatively large active thicknesses (at least 0.1 μm in a double heterostructure) as the volume of the active region is decreased, the current necessary to achieve threshold decreases. A major emphasis then is to decrease the thickness of the active region in laser diode design.

Figure 5 shows a schematic of the transverse radiation distribution of electromagnetic energy for a single-mode laser. Since much of the energy in a laser diode travels in the cladding regions, these regions must be included in analysis of the laser. Only a fraction, Γ, of the radiation in Fig. 5 is within the active region. Thus a wavefront propagating within the laser cavity sees an average absorption coefficient $\langle \alpha \rangle$ that is influenced by the cladding:

$$\langle \alpha \rangle = (1 - \Gamma)\alpha_{\text{clad}} + \Gamma\alpha_{\text{active}} \tag{1}$$

where α_{clad} and α_{active} are the absorption coefficients of the cladding and

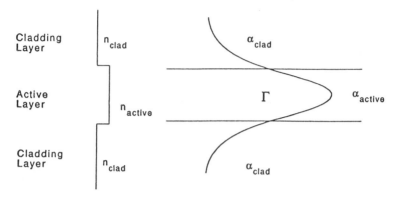

Figure 5   Description of the differing loss coefficients α_{active} and α_{clad}.

the active regions, respectively. If we define a gain coefficient g for the device, we can write an equation for the intensity of a wave as it travels down the guide in the z direction:

$$I(z) = I(0)\exp[(\Gamma g - \langle\alpha\rangle)z] \tag{2}$$

where the gain coefficient and the absorption coefficients are both functions of the semiconductor composition and photon energy. The lasing threshold condition is met when the round-trip gain in the laser exceeds the losses. For a cavity of length L we can write Eq. 2 as

$$I(2L) = I(0)R_1R_2 \exp[(\Gamma g - \langle\alpha\rangle)2L] \tag{3}$$

where R_1 and R_2 are the reflection coefficients of the two end mirrors. In steady state, the lasing condition is met when the magnitude and phase of the returning wave equal that of the original wave. For the magnitude of the wave, this means

$$I(2L) = I(0) \tag{4}$$

For the phase

$$e^{-i2\beta L} = 1 \tag{5}$$

where β is the propagation constant equal to $2\pi/\lambda n$. Equations 4 and 5 tell us which vibrational modes will have sufficient gain to achieve oscillation.

The condition for reaching the lasing threshold is determined by the optical gain threshold g_{th} found from solving Eq. 3 at the point where $I(z) = I(2L)$, which is the point where the gain equals the loss:

$$\Gamma g_{th} = \langle \alpha \rangle + \frac{1}{2L} \ln\left(\frac{1}{R_1 R_2}\right) \tag{6}$$

The resonance vibrational frequency is determined by Eq. 5 as holding when

$$\beta = \frac{\pi m}{L} \tag{7}$$

where m is an integer. We can determine the vibrational frequency of each mode in the cavity as

$$\nu_m = \frac{mc\pi}{2Ln} \tag{8}$$

Most laser diodes will have multiple longitudinal modes, although as discussed, some may be specifically designed to oscillate at the fundamental.

2.1 Threshold Gain

The lasing threshold gain can be determined from knowledge of the stimulated and spontaneous recombination-rate parameters derived in Section 3.1 in Chapter 2. In that section, the rate parameters were derived considering recombination of e-h pair under equilibrium conditions. We will modify this for the conditions illustrated in Fig. 1, that is, the system is under heavy current injection, and the equilibrium Fermi level must be replaced by the respective quasi-Fermi Levels E_{Fn} and E_{Fp}. These levels describe, respectively, the population densities in the conduction and valence band under the conditions of heavy injection which exist at threshold. We first rewrite the relation between the spontaneous (r_{sp}) and stimulated emission rate (r_{st}) according to Eq. 58 of Chapter 2

$$r_{st}(E) = r_{sp}(E) \left\{ 1 - \exp\left[\frac{E - \Delta E_F}{kT}\right] \right\} \tag{9}$$

where $\Delta E_F = E_{Fn} - E_{Fp}$ and $E = h\nu$. Here r_{sp} and r_{st} are, respectively, the spontaneous and stimulated rates per unit volume. This equation is derived in the same manner as Eq. 58 of Chapter 2, except that consideration of the nonequilibrium representation of Fig. 3 is now included. We can now observe that the stimulated emission will only predominate when $h\nu \ll \Delta E_F$. If the mode density per unit photon energy is given by [2]

$$p(E) = \frac{8\pi n^3 E^2}{c^3 h^3} \quad (\text{erg}^{-1} \text{ cm}^{-3}) \tag{10}$$

The power emitted per unit volume can be written as $NEr_{st}(E)$, where N

is the photon density per mode. The power crossing the cavity per unit area can be written as $vENp(E)$, where v is the velocity of light (c/n). The gain can be written then as

$$g(E) = \frac{NEr_{st}(E)}{vENp(E)} = \frac{r_{st}}{vp(E)} \qquad (11)$$

which yields

$$g(E) = \frac{c^2 h^3}{8\pi n^2 E^2} r_{sp}(E) \left[1 - \exp\left(\frac{E - \Delta E_F}{kT}\right) \right] \qquad (12)$$

In the above formulation, we can assume band-to-band recombination and use an approach similar to Eq. 59 of Chapter 2 to determine r_{sp}. A more rigorous model would include modeling the absorption within and above the band gap. One could, for example, evaluate the matrix elements for upper to lower transition states with a density-of-states model that includes band tails. This is most commonly attributed to Stern [3], who considered the transition rate at photon energy $h\nu$ between a density of vacant valence band states $\rho_v(E)$ and occupied conduction band states $\rho_c(E + h\nu)$ as

$$g(h\nu) = \frac{q^2}{2m_0^2 \epsilon_0 ncv} \int_{-\infty}^{\infty} \rho_c(E + h\nu)\rho_v(E)|M(E + h\nu)|^2$$

$$X[f_v(E) - f_c(E + h\nu)] \, dE \qquad (13)$$

where $|M(E + h\nu)|$ is the matrix element representing individual transitions and f_v and f_c represent the quasi-Fermi levels which now reside in the valence and conduction bands, respectfully. The density of states function can be replaced by density of band tail states to make the representation more accurate.

Equation 13 is negative (meaning net absorption instead of net gain) for $f_v(E) < f_c(E + h\nu)$. This means there is only net gain for $E_{Fc} - E_{Fv} > h\nu$, so only light of energy less than the separation of the quasi-Fermi level energies experience net gain. The general shape of the gain curve is shown in Fig. 6 [4]. The gain peaks between $h\nu = E_{Fn} - E_{Fp}$ and $h\nu = E_g$, below which no electronic transitions can occur (barring tail states). Figure 6 shows calculated plots based on parameters for GaAs. The figure shows similar dependences on injected carrier densities. When the density is low, the separation between the quasi-Fermi levels is less, and at some value of this separation the gain will be zero at $E_{Fn} - E_{Fp} = h\nu$.

Figure 7 [5] shows calculated gain as a function of injected carrier density for a GaInAsP laser diode at various temperatures. The gain coeffi-

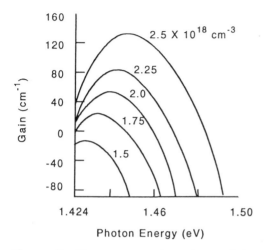

Figure 6 Photon energy dependence of the optical gain (or loss) of GaAs at different injected carrier densities. (From Ref. 4.)

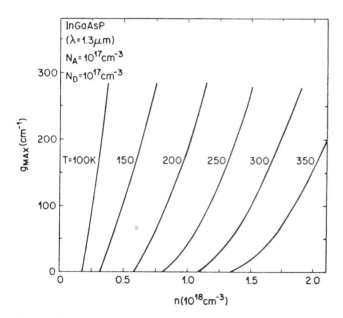

Figure 7 Maximum gain as a function of injected carrier density for undoped $\lambda = 1.3$ μm GaInAsP at different temperatures [5].

cient as a function of the injected carrier density can be approximated by

$$g \propto (n - n_{trans})^b \tag{14}$$

where n_{trans} is the minimum pair density needed to produce population inversion, i.e., where the gain is zero, or the material is "transparent." The value of n_{trans} for GaAs based on Fig. 7 is ~1.55 × 10^{18} cm^{-3}. The parameter b is a constant that varies between 1 and 3, depending on material, temperature, and gain value. The result of the gain threshold, and the dependence of Eq. 14, is a threshold current below which only spontaneous emission is obtained. Above this threshold, when the pair carrier density is such that $n > n_{trans}$, then the gain factor exceeds losses and growth of a particular vibrational mode (or modes) is obtained at the expense of the other modes. Which mode exceeds threshold first is dependent on Eqs. 4 and 5.

From Fig. 7, it appears that very high gain factors (up to several hundred cm^{-1}) are possible in laser diodes. In an actual device, the gain is limited by *gain saturation*. If the photon density is high enough, the time needed to replenish the conduction band population becomes comparable to the carrier lifetime. Up transition rate no longer exceeds recombination rate, and the gain saturates. Gain factors of 80 cm^{-1} are typical in GaAs devices. High gain factors in laser diodes as compared with other laser technologies are a result of the density of states being dependent on Fermi-Dirac instead of Boltzmann statistics.

We can write an equation for the gain factor equation 14 as a function of current density:

$$g = \frac{\beta \eta_i}{d} (J - J_0)^b \tag{15}$$

where β and J_0 are constants that depend on the laser structure and composition of the active layer, and η_i is the internal quantum efficiency. If we let $b = 1$ then, at threshold,

$$J_{th} = \frac{g_{th} d}{\beta \eta_i} + J_0 \tag{16}$$

$$J_{th} = \frac{d}{\beta \Gamma \eta_i} \left(\langle \alpha \rangle + \frac{1}{2L} \ln\left(\frac{1}{R_1 R_2}\right) \right) + J_0 \tag{17}$$

Figure 8 [6] shows variation of J_{th} with cavity length and active layer thickness for various AlGaAs compositions. To a first approximation, the threshold current is linearly proportional to the active layer thickness.

Both the measured constants β and J_0 are temperature dependent. As the temperature rises, the amount of absorption increases, and the gain

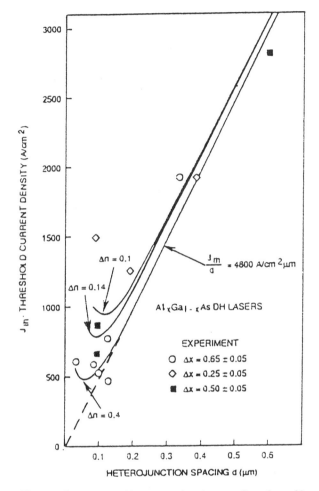

Figure 8 Threshold current density as a function of heterojunction spacing [6].

necessary to achieve threshold, g_{th} increases. J_{th} also increases, β decreases, and J_0 increases with increasing temperature [7]. The temperature dependence of J_{th} is often modeled as

$$J_{th} \propto \exp\left(\frac{T}{T_0}\right) \tag{18}$$

where $T_0 = 160$ K for GaAs for $0° \leq T \leq 100°$C. Observed variations of

J_{th} with temperature are shown in Fig. 9 [8]. Also show is data from older diode designs, single and homojunction laser diodes.

As the injection current is raised to the point of threshold, the optical gain begins increasing, and the refractive index begins decreasing. This is observable in the real part, n', and the imaginary part, n'' of the refractive index, both of which are related through the Kramers-Krönig relations:

$$\Delta n'(E) = \frac{2}{\pi} P \int \frac{E' \, \Delta n''(E') \, dE'}{E'^2 - E^2} \tag{19}$$

where $\Delta n = \Delta n' + i \, \Delta n''$ represents the change in the refractive index with injection current, and P is the principle value of the integral. The imaginary part of the refractive index is related to the change in gain and absorption in the material, and thus is related to the gain of the laser:

$$\Delta g = -2 \left(\frac{\omega}{c}\right) \Delta n'' \tag{20}$$

the propagating wavevector, k, is complex and is related to the complex dielectric constant:

$$k = k' + ik'' = \left(\frac{\omega}{c}\right) n = \left(\frac{\omega}{c}\right) \epsilon^{1/2} \tag{21}$$

where $\epsilon = \epsilon' + \epsilon''$.

Figure 10 [9] shows the variation of the real and imaginary parts of the refractive index as the laser approaches threshold. An important laser

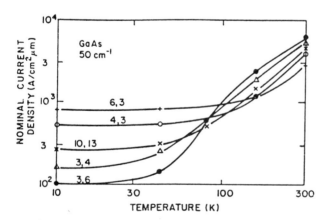

Figure 9 Current density vs. temperature for various donor and acceptor concentrations. (From Ref. 8, © 1973 IEEE.)

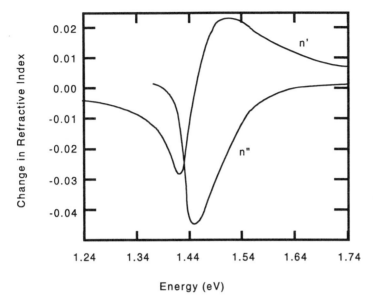

Figure 10 Changes in the real and imaginary parts of the refractive index of a GaAs laser when it is excited from low current up to the threshold current. (From Ref. 9.)

parameter is the linewidth enhancement factor, α, defined by

$$\alpha \equiv \frac{\partial n'/\partial N}{\partial n''/\partial N} = \frac{\Delta n'}{\Delta n''} \tag{22}$$

where N is the injected current density. In order to obtain a fine linewidth, it is necessary to operate the laser on the short-wavelength side of the gain peak of Fig. 6, where the linewidth enhancement factor is smaller. As the emission wavelength increases, the linewidth increases. Figure 11 [10] shows linewidth enhancement factors measured for GaInAsP.

2.2 Quantum Efficiency

The differential quantum efficiency can be defined as

$$\eta_{\text{ext}} = \frac{\text{increase in the number of photons emitted}}{\text{increase in carrier pair density above threshold}}$$

$$= \frac{dP}{d[(I - I_{\text{th}})h\nu/e]} \tag{23}$$

where P is the output power, and thus can be found from the derivative

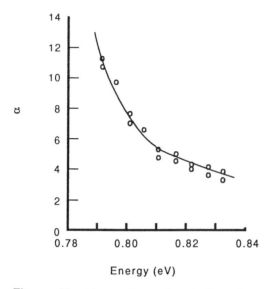

Figure 11 Measured wavelength dependence of the linewidth enhancement factor for $\lambda = 1.5 \ \mu m$ GaInAsP laser diodes. (From Ref. 10.)

of the dependence of P on I_{th}. Power output, P, is the product of the stimulated emission power

$$\frac{(I - I_{th})\eta_i}{e} \, h\nu \tag{24}$$

and a factor representing the photon escape probability for stimulated emission:

$$\frac{(1/L)\ln(1/R)}{\langle\alpha\rangle + (1/L)\ln(1/R)} \tag{25}$$

where η_i is the internal quantum efficiency. The factor $(1/L)\ln(1/R)$ represents the output loss of the mirror. Equation 23 then yields

$$\eta_{ext} = \frac{\eta_i}{1 + \langle\alpha\rangle L/\ln(1/R)} \tag{26}$$

By plotting η_{ext} versus $1/L$ for diodes of various lengths and extrapolating to $L = 0$, the internal quantum efficiency can be determined. For GaAs, η_i is usually found to be ~ 0.9.

2.3 Laser Diode Modulation

It is most fortunate that laser diodes can be easily modulated at very high frequencies. Modulation of laser diodes is of interest for applications in digital data processing and high-speed communications. Unlike other laser systems, laser diodes can be modulated without seriously affecting beam quality or lasing conditions. This is done simply by modulating the drive current to the device. Modulating drive current modulates the gain of the laser, which in turn modulates the output power.

Frequency response characteristics have been analyzed by several authors, and the response is found to have a characteristic shape. For low frequencies, the response is very flat but peaks at high frequencies at a resonance or relaxation oscillation frequency f_r. The value of f_r is found in all cases to be sensitive to the slope of the line of gain versus injected current density. We recall Eq. 14 as

$$g = B(n - n_{\text{trans}}) \tag{27}$$

where n is the injected current density and n_{trans} is the "transparent" density, which is the minimum density necessary for population inversion. In this equation, the proportionality constant B is the slope of the gain-versus-pair density curve.

If the photon density inside the active region is given as $P(\nu)$, the change in density with time is equal to the generation rate of photons minus the photon lifetime. The generation rate of stimulated transitions per unit volume is given by $A(n - n_{\text{trans}})P(\nu)$, where A is a constant. The change in photon density with time in the cavity is given by

$$\frac{dP(\nu)}{dt} = A(n - n_{\text{trans}})P\Gamma - \frac{P}{\tau_P} \tag{28}$$

where Γ is the confinement factor and τ_P is the photon lifetime. At equilibrium, Eq. 28 equals zero and

$$A(n - n_{\text{trans}}) = \frac{1}{\Gamma \tau_P} \tag{29}$$

but the photon lifetime in the active region is related to the gain by

$$\frac{1}{\Gamma \tau_P} = g \frac{c}{n_r} \tag{30}$$

where n_r is now the refractive index in the active region. This results in a relation between B and A given by

$$A = \frac{Bc}{n_r} \tag{31}$$

The change in injected carrier density with time is given by

$$\frac{dn}{dt} = \frac{J}{eL} - A(n - n_{\text{trans}})P\Gamma - \frac{n}{\tau_n} \tag{32}$$

where L is the active layer thickness and τ_n is the recombination lifetime. In this analysis, we have ignored the contribution of spontaneous emission, since this represents only a small fraction of the lasing mode.

If the injected current is given by a DC plus AC current:

$$J = J_0 + J_1 e^{i\omega t} \tag{33}$$

and the pair density n and optical power density P are both similarly modulated,

$$n = n_0 + n_1 e^{i\omega t}, \qquad P = P_0 + P_1 e^{i\omega t} \tag{34}$$

It has been shown in Ref. 11 that the resonance frequency is

$$f_r = \frac{1}{2\pi} \sqrt{\frac{n_r B P_0}{c \tau_p}} \tag{35}$$

From this result, we see that the modulation bandwidth of the laser diode is dependent upon (1) the differential gain coefficient, B, (2) photon density, and (3) photon lifetime.

The modulation frequency can be increased according to Eq. 35, by increasing the differential gain B (which equals dg/dn). This can be done by quantum confinement (see Section 7) and by cooling the laser. The photon density can be increased by increasing the confinement of photons, that is, by maximizing the waveguide properties of the active region. And finally, the photon lifetime can be decreased simply by shortening the cavity or by increasing the drive current. Each of these have been implemented with varying success.

3 TRANSVERSE CONFIGURATIONS: SEPARATE CONFINEMENT, AND GRADED INDEX CHANNELS

The active volume—that is, the total volume of the waveguide—is normally smaller than the mode volume, where the vibrational modes are confined. This is because much of the electromagnetic radiation travels in the cladding. As the guide thickness of the laser is increased, less radiation travels in the cladding, the spot size illuminating the exit facet becomes smaller, and the threshold current increases according to Eq. 17. Consequently, the maximum possible output power decreases, as possibility of facet damage from a small spot size is high.

The laser can also purposely be constructed so that the mode volume can be adjusted separately from the active volume. Figure 12 shows how refractive index variations in the transverse direction can be grown into the diode. The active region is still defined by the "kink," or smaller-band-gap region, but the mode volume is adjustable separately from this kink by adjusting the thickness of a separate waveguide region. If the active volume is made very thin (as thin as ~0.05 μm) then most of the optical energy will spread into the guide layer. This acts to increase the spot size of the laser, which in turn reduces the power density which illuminates the facets and lowers the possibility of facet damage. As shown by Eq. 17 in Section 2, this also allows separate control of the threshold current for thin active layers. Higher continuous output powers are then possible. A separate confinement laser created for this purpose is often termed a *large optical cavity* (LOC) laser. Thinning the active layer from 0.15 to 0.05 μm nearly doubles the transverse size of the lasing spot from 0.6 to 1.0 μm [12]. Addition of the optical cavity in a LOC laser increases the spot size an additional 50–70%.

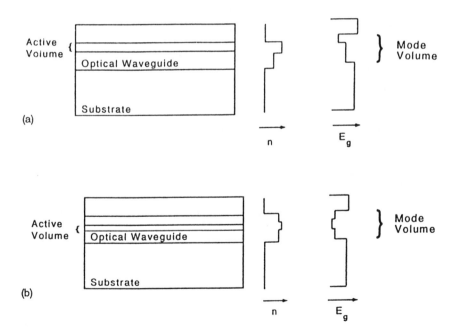

Figure 12 Examples of (a) large optical cavity (LOC) and (b) separate confinement heterostructure (SCH) transverse configurations.

LOC lasers can also be used in distributed feedback lasers (discussed in Section 5.4) to separate the active region from the corrugations in the guide, where a large amount of nonradiative recombination is experienced.

A separate confinement heterostructure laser diode is similar to the LOC device. In SCH lasers, a separate optical wavguide layer exists both above and below the active region (Fig. 12b).

4 LATERAL CONFIGURATIONS

While the double-heterostructure design of Fig. 3c can confine carriers and act as a waveguide in the transverse direction, confinement in the lateral direction is also desired. This can be a difficult task: in order to minimize facet damage the emitted beam must be spread out in the lateral dimension to incident the facet as uniformly as possible. The "sides" in the lateral dimension must then act as a waveguide with the active region, confining the light to this region. Three possible lateral confinement methods include gain guiding, positive-index guiding, and negative-index guiding. These three structures are represented in Figs. 13 through 15.

4.1 Gain Guided

In a gain-guided, or stripe contact laser, electrical contact to the upper semiconductor surface is made by way of a metal stripe, the purpose of which is to confine the injected current into an area determined by the dimensions of the stripe. This contact may be made quite thick, and the device mounted upside down on the stripe provides heat dissipation. As there is actually a modest decrease of the refracted index due to the injected current, there is no lateral waveguide in the dielectric sense, and the only way to contain the emission is by control of the lateral spread of the current. These lasers may provide very high output power; however, the lateral spread of the injected current is considerable and guiding is very weak. The result is a large spread in the lateral modes, and the far-field pattern of gain-guided lasers will be broad and may actually have two peaks.

To improve current confinement in a stripe contact laser, confinement regions may be introduced into the material. Figure 13b shows enhanced current confinement through ion implantation. Implantation of protons (H^+) creates damaged regions outside the laser guide; these damaged regions are semi-insulating, that is, possess a resistivity of $\sim 10^7$ Ω-cm (see Section 7 in Chapter 1) and confine current to the active region. Alternatively, an area within an n-type cap can be diffused with a p-type

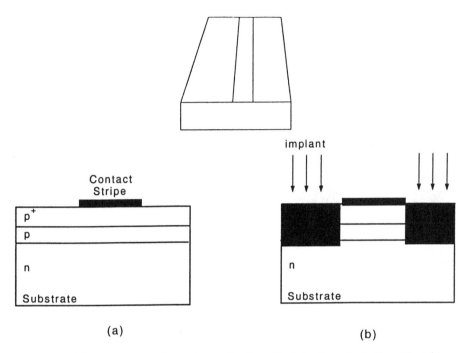

Figure 13 Lateral configurations of gain-guided devices, including (a) stripe contact and (b) implant isolated laser diodes.

dopant, and current is then confined in the lateral dimension by reversed-bias *p-n* junctions.

Lateral blocking layers can also be introduced by epitaxial growth of *p-n* junctions built into the devices. Figure 14 shows a double-heterostructure device with an *n*-type blocking layer. Note that the device is made on *p*-type substrates instead of *n*-type. The *n*-type blocking layer forms

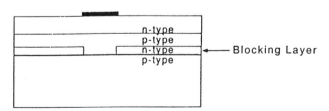

Figure 14 Improvement in current confinement in a gain-guided device by addition of an *n*-type blocking layer in an *n-p-n-p* device.

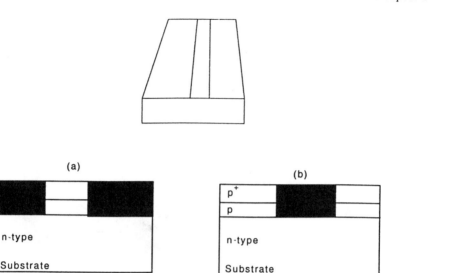

Figure 15 Index-guided lateral configurations, including (a) positive and (b) negative-index-guided devices.

part of a *p-n-p* structure, confining the current to an opening of the blocking layer. If a *n-p-n* type structure were to be used on an *n*-type substrate, electrons would diffuse right through the *p*-type blocking layer because of the long minority carrier diffusion length for electrons. Thus, a relatively thick *p*-type blocking layer would be required. Since holes have a much shorter diffusion length than electrons, a much thinner blocking layer can be used in the *p-n-p* structure.

4.2 Index Guided

Positive- and negative-index guided lasers have physical structures which provide real dielectric lateral-index waveguiding. The increase (or decrease) in the refractive index to form the central region of the guide can be achieved in a variety of ways, and the results are summarized in Fig. 15. In each case modes in the lateral dimension are stripped either by absorption, as in positive waveguides, or by refraction, for a negative-index waveguide. Higher modes thus penetrate the lossy region surrounding the guide more than the fundamental and thus have a higher threshold condition for lasing. Index-guided lasers are more complex and require more intricate processing and are therefore more costly than gain-guided devices.

In the positive-index guide, the refractive index of the material inside the guide is higher than that outside the guide. The relative difference in the refractive index and the width of the guide are chosen so that only the fundamental mode is supported. A single-mode diode laser is one that will support only the fundamental vibrational mode in the transverse direction and only the fundamental vibrational mode in the lateral dimension as well. The output is a Gaussian profile more collimated in the far-field pattern than the less expensive gain guided laser.

In a negative-guided index laser (the "antiguide" laser) the refractive index of the guide is lower than the surrounding medium, and the guide acts to strip unwanted modes which have higher losses than the fundamental. Fundamental vibrational mode is then supportable.

The buried heterostructure is a popular choice for index-guided lasers. For this structure, the epitaxial layers which make up the double heterostructure are first prepared as in the gain-guided device. Then a mesa is etched which gives the buried heterostructure its characteristic shape. The wafer is then returned to the epitaxial growth chamber, and the remainder of the material is regrown to "bury" the waveguide. Both positive- and negative-index devices can be formed by this method.

In the selectively diffused index-guided laser, a dopant is diffused into the device in the area where the index change is desired. A dopant is selected that will raise (or lower) the refractive index of the guide. Refractive index changes are then a direct result of the presence of the dopant.

4.3 Bent Waveguides

It is also possible to obtain index guiding without additional processing steps inserted between the sequence of epitaxial growths. Such structures, shown in Fig. 16, include the channeled-substrate planar (CSP) and the

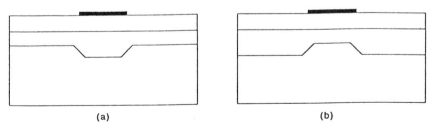

(a)   (b)

Figure 16 Examples of bent waveguide index-guided lateral configurations, including (a) channel substrate planar and (b) constricted double heterostructure.

constricted double heterostructure (CDH), During growth, the channel in the substrate is filled in, creating variations in thickness. The optical field perceives a thicker area to be a positive-index guide. In the CDH laser, the light perceives bends in the active region as positive-index guides.

5 LONGITUDINAL CONFIGURATIONS

5.1 Fabry-Perot Cavity Lasers

A traditional Fabry-Perot cavity can be formed in a diode laser by cleaving the ends of the laser in the longitudinal dimension. Because of the high-refractive-index change between the cleaved ends and the outside, a high reflectivity coefficient ($\sim$0.3 for GaAs) can be obtained. In addition, to lower reflectivity and thus reduce the possibility of facet damage, antireflection coatings may be applied to one or both ends. Typical cavity sizes may be 200–300 μm (longitudinal) long, 10 μm wide (lateral), and 200 nm thick (transverse). Vibrational modes can be obtained in all directions. Lateral modes are sometimes discouraged by rough-cutting the edges of the diode so constructive interference is less likely. Lateral modes can also be controlled by creating dielectric waveguides in the lateral direction as discussed previously.

For a diode of length L, the separation between individual longitudinal modes can be determined from $L = m\lambda/2n$ as

$$\frac{\Delta\lambda}{\Delta m} = \frac{\lambda^2}{2L[(dn/d\lambda)\lambda - n]} \tag{36}$$

where n is the effective refractive index of the propagating mode. If the variation of n with wavelength $dn/d\lambda$ is small, then

$$\frac{\Delta\lambda}{\Delta m} = \frac{-\lambda^2}{2Ln} \tag{37}$$

Figure 17 [13] shows the mode spacing of double-heterojunction AlGaAs laser diodes at various current levels.

5.2 Dynamic Single-Mode Operation

While single-mode operation can be obtained for the transverse and longitudinal configurations, dynamic single-mode (DSM) operation for the longitudinal configuration is also desirable. Reducing the number of modes results in minimizing modal dispersion in fiber optics and thus reduces pulse spreading. Reducing the mode number also reduces partition noise encountered in high-speed modulation.

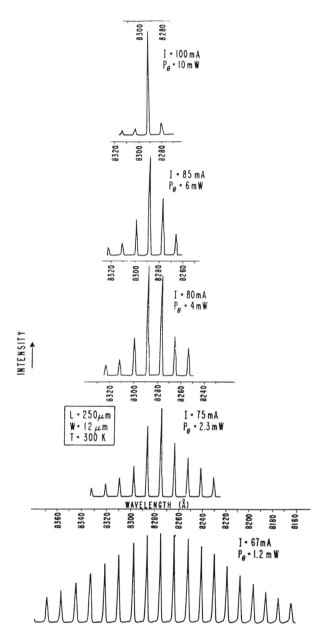

Figure 17 Lasing spectrum from a AlGaAs/GaAs double-heterostructure laser diode at various current levels [13].

For index-guided lasers, single-mode operation is possible but only under special conditions. If the current is modulated, the slight variations of refractive index with population density cause many modes to appear, and thus fundamental oscillation is easiest in DC operation. To achieve single-mode operation in pulsed-mode form, there are four alternatives:

1. Short cavities
2. Cleaved-coupled cavities
3. External energy injection
4. Distributed feedback

Short Cavities

When the cavity length is short, according to Eq. 37, the mode spacing is large and multimode oscillation is less likely because the short cavity cannot support extra modes. Furthermore, if the mode spacing is large, the oscillator is less likely to "jump" from one mode to another. Short-cavity lasers can be prepared as short as 40–60 μm by microcleaving (compared to normal ~400 μm). However, as the length is shortened, the threshold for lasing for all modes is increased, including the fundamental, and the threshold current increases. Furthermore, there are immense difficulties in handling short-cavity lasers.

By reducing the cavity length to 25 μm for a conventional Fabry-Perot cavity, the mode spacing can be increased to ~10 nm for a laser operating at 1.3-μm wavelength. Since the losses for each mode are nearly identical, the gain curve for diode lasers is broad enough that any narrower mode spacing will result in multimode oscillation. Mode discrimination is realized by centering on one mode through temperature or wavelength tuning.

Conventional cleaved mirror Fabry-Perot structures are difficult to fabricate at resonator lengths less than 50 μm. Microlasers have been made as short as 20 μm by reactive ion etching [14]. Other microcleaved devices with side-mode powers suppressed by several orders of magnitude have been fabricated, but most short cleaved structures are difficult to fabricate and reproduce.

An alternative to the conventional cleaved mirror laser is to fabricate a vertical device similar to that shown in Fig. 18 [15]. This figure is of a vertical cavity–surface emitting laser. Here the resonator, of length 6 μm, is formed in the direction normal to the surface of the substrate and the light exits the top instead of the side. A selective growth-regrowth process is first used to prepare the buried heterostructure. One reflector is formed from a layer of gold and multilayers of SiO_2/TiO_2 which are evaporated on the surface of the device. A hole is etched in the substrate, and the other SiO_2/TiO_2 partial reflector is made within this hole. The transverse dimension is circular of 10 μm diameter. The spectral width above thresh-

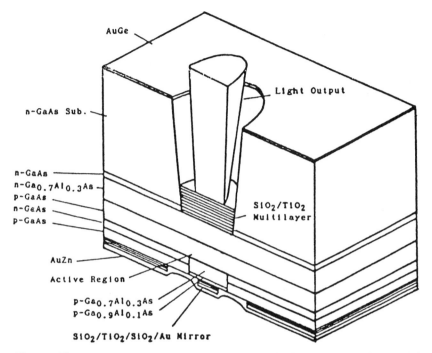

Figure 18 Schematic of a vertical-cavity, dynamic single-mode surface emitting laser. (From Ref. 15, © 1992 IEEE.)

old was less than 0.1 nm, which was the limit of the measuring spectrometer. The mode space within the device was a full 17 μm.

Having a surface emitting device holds several advantages in processing and functionality. The laser can be fabricated from a fully monolithic process and is only separated into chips before final packaging. This also facilitates on-wafer probe testing. Initial probe testing of conventional cleaved devices before separation into individual chips is not possible. Since the light does not exit the side, vertical emitters are more favorable for integration into optoelectronic integrated circuits.

Cleaved-Coupled Cavity Lasers

If a narrow groove is etched in the laser perpendicular to the stripe of Fig. 13, one side of the groove would act as a mirror for the longitudinal cavity and the other side of the groove would act as an "external" mirror for the same cavity. Thus a three-mirror resonator is formed. Single-longitudinal-mode selection can take place with the external mirror by the coupled-cavity effect. Since the optical path lengths from the two reflections are not the same, the modes of the double cavity are widely spaced. The current and temperature of the device must be carefully controlled in order to select the desired mode. Coupling can be enhanced in a three-mirror resonator using a graded index (GRIN) lens as the external mirror (Fig. 19 [16]).

A four-wave resonator can be made by cleaving the cavity. A gap is formed by cleaving a chip and slightly separating the two devices [16]. The two sections can be driven separately and with lasing in both sections;

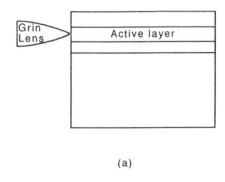

(a)

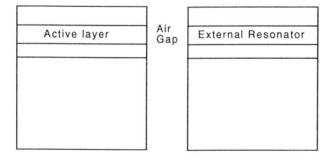

(b)

Figure 19 Multiple-element lasers, including (a) external GRIN-rod coupled cavity and (b) four mirror cleaved-coupled cavity (C³) configurations. (From Ref. 16.)

a gap of $\lambda/2$ provides optimum selectivity. If one section is driven above threshold, the other is used as a tuning element which controls the refractive index (and hence the effective length of the cavity) through variation of the injected current density in the tuning element. Coupled cavity, with the combination of short cavity, has been used to observe single-mode operation [17].

5.3 External Light Injection

A laser diode can also be mode-locked by pumping from an external laser. The oscillation wavelength of the diode can be tuned to the frequency of the pump and the longitudinal mode stabilized. In addition, the linewidth of the laser diode output can be determined by the linewidth of the exciting source. The 1-GHz linewidth of an isolated laser diode was narrowed to 15 MHz by He-Ne laser injection locking [17].

5.4 Distributed Feedback

Diode lasers have flexibility in that a Fabry-Perot cavity is not the only feedback structure available for use. A distributed feedback configuration creates a laser that is frequency selective and, for certain feedback types, surface emitting as well.

In a distributed feedback laser (DFB), shown in Fig. 20, a corrugation is etched in the waveguide section of the large optical cavity laser. This pattern is made using lithography by illumination of a interferometric pattern. The wafer is then returned to the epitaxial growth chamber, and the remainder of the epitaxial layers are prepared to complete the structure. The grating acts as a Bragg diffraction grating which provides feedback for light. The general theory for Bragg diffraction can be obtained from Fig. 21. In this figure, rays 1 and 2 are diffracted from the incident light. Perpendicular lines are drawn from scattered rays 1 and 2, and the distance between these lines is b. The condition for constructive interference between diffracted waves 1 and 2 is met when

$$b + \Lambda = m\left(\frac{\lambda_0}{n}\right), \qquad m = 0,1,2,\ 3,\ \ldots$$

where Λ is the grating wavelength. Since $b = \Lambda \sin\theta$,

$$\sin\theta = \frac{m\lambda_0}{n\Lambda} - 1 \tag{38}$$

In this case the incident light is generated in the plane of the junction. Thus, the incident angle is 0° and the Bragg condition is

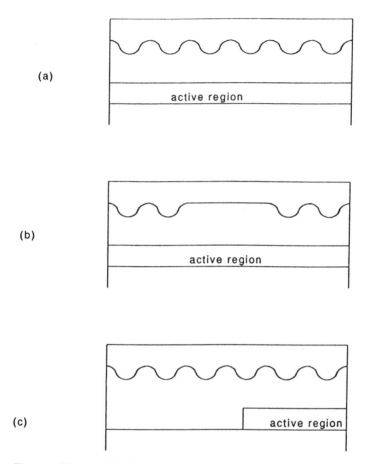

Figure 20 (a) Distributed feedback (DFB), (b) distributed Bragg reflected (DBR), and (c) distributed reflected (DR) longitudinal configurations.

$$\Lambda = p \left(\frac{\lambda_B}{2n} \right), \qquad p = 1,2,3, \ldots \tag{39}$$

where λ_B is the Bragg wavelength. Substituting this into Eq. 38 gives for the diffracted angle

$$\sin \theta = \frac{2m}{p} - 1 \tag{40}$$

which restricts m to $0 \leq m \leq p$ for θ to be real. We note from the figure

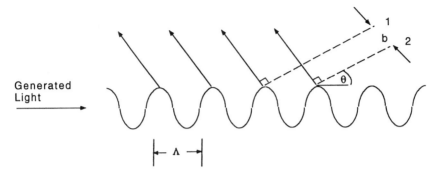

Figure 21 General principles of distributed feedback.

that if any light were to exit the front surface of the device, the scattered light from the grating would be incident on the surface of the semiconductor at an angle θ.

For the first-order Bragg condition, $p = 1$ and $m = 0$. The angle $\theta = -90°$, and the light is scattered in the forward direction. For the second order, $p = 2$ and $m = 0$ also results in forward scattering. But for $p = 1$, $m = 1$, the scattered angle $\theta = 90°$ and the ray is scattered in the backward direction for distributed feedback.

Figure 22 summarizes the possible scattering angles for the first three orders. We note that for higher orders there exist several possible values of m for which the light can be scattered toward the surface. However, since only light within the critical angle (e.g., $\theta \leq 16°$ for GaAs) will be able to exit the surface, surface emitting lasers based on Bragg reflection are limited to even values of p. Thus, surface emitting lasers can be prepared from even grating orders and nonsurface emitting devices from odd grating orders. The exit angle ϕ from the surface of the semiconductor is determined by Snell's law and is equal to

$$\sin \phi = \frac{m\lambda}{\Lambda} - n \tag{41}$$

where the angle ϕ is measured from the perpendicular of the surface. The output beam divergence can be determined from differentiating Eq. 41:

$$d\phi = \frac{\Delta\lambda n}{\lambda \cos \phi}\left[\frac{m\lambda}{n\Lambda} - \frac{\lambda}{n}\frac{dn}{d\lambda}\right] \tag{42}$$

If the bandwidth $\Delta\lambda$ is 6 Å, the angular spread is 0.21°. Added to this is the divergence normal to a 500-μm radiator, ~0.1°.

1st Order

$$\Lambda = \frac{\lambda_0}{2n_{eff}}$$

$m=1 \qquad m=0$

2nd Order

$$\Lambda = \frac{\lambda_0}{n_{eff}}$$

$m=1$

$m=2 \qquad m=0$

3rd Order

$$\Lambda = \frac{3\lambda_0}{2n_{eff}}$$

$m=2 \qquad m=1$

$m=3 \qquad m=0$

Figure 22   Possible Bragg scattering for the first three orders.

The distributed feedback laser benefits from a lower sensitivity to drive current and operating temperature than the conventional Fabry-Perot design, since temperature affects only the index of refraction and is not sensitive to band-gap changes as in the conventional design.

It was soon discovered that if the corrugation were made at the interface between the cladding and active regions, a large amount of nonradiative recombination would result disrupting lasing action. To alleviate this problem, a large optical cavity or separate confinement heterostructure (Section 3) is used to separate the active region from the feedback region (separate confinement distributed feedback, or SC-DFB).

In the distributed Bragg reflector (DBR) the corrugation exists only at each end of the laser, and reflection from these areas is much like the mirrors of the cavity lasers. The distributed reflector (DR) laser is a combination of DFB and DBR lasers combining some passive feedback with active. Figure 20 shows schematics of these different configurations.

In an ideal distributed feedback laser, the longitudinal vibrational modes are separated about λ_B as

$$\lambda = \lambda_B \pm \frac{\lambda_B^2}{2nL_\theta} \left(q + \frac{1}{2} \right) \tag{43}$$

where q is the mode order (starting at 0) and L_θ is the effective grating length. Although many possible longitudinal modes exist, the two for $q = 0$ will have the lowest threshold gain, and one of these will dominate due to the randomness of the cleaving process. The asymmetry of the facet reflections lifts the degeneracy of the two $q = 0$ modes and one will dominate. This asymmetry can be increased by applying unequal antireflection coatings (e.g., 2% on one face, 30% on the other) on each facet.

6 LASER ARRAYS

To achieve high output power, 100 mW to possibly several watts, arrays of phase-locked diode lasers must be used. Eight, 10, or sometimes as many as 20 laser diodes can be fabricated monolithically on a single substrate. If the devices are close together, optical modes which spread into the cladding of each laser begin to interact with one another. The effect is coupling between these modes, which causes the laser outputs to couple. It is desirable to couple as many devices in an array as possible; increasing the number of devices not only raises the output power but also raises the effective output aperture. The output beam divergence is then reduced and a more collimated laser spot size results.

The coupling between devices in an array is complex. There must be zero phase shift between each laser element in the array in order to obtain a single-lobed beam in the far field. Any other phase shift can cause multiple lobes. Thus, lateral mode control is an important issue. High-order modes also cause multiple lobes which must be π-phase shifted to achieve a single well-defined beam.

In a gain-guided device, there is no lateral optical confinement, and a variety of complex multimodes exist. These modes are relatively unstable due to dependences on drive current and junction temperature. The far-field pattern is a complex multilobed beam which is difficult to couple into optical systems. Thus, the most common method of stabilizing lateral modes between elements in the array is to use index-guided lateral configurations.

In a positive-index guide (Section 4.2) high-order modes within the active region are stripped out by absorption. Thus, coupling between posi-

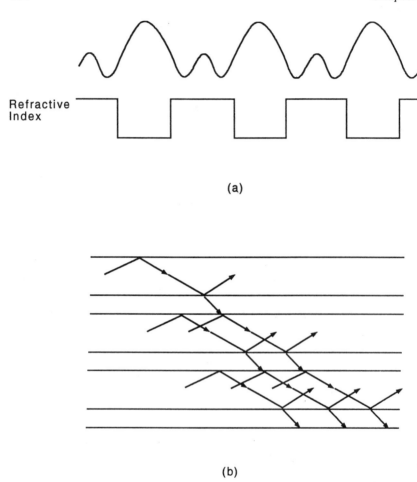

Refractive
Index

(a)

(b)

Figure 23   Schematic of (a) resonant coupling conditions in antiguided array, (b) ray-optics representation of a resonant optical waveguide (row) [18].

tive-index devices in an array exist from the evanescent modes in the cladding regions. Because evanescent modes are weak, higher drive currents are then necessary for effective intermodal coupling, and the possibility of spatial hole burning of the fundamental mode is possible, resulting in self-focusing of that modes' output.

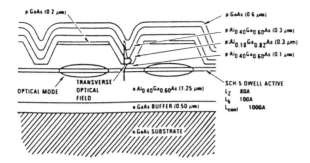

Figure 24 Schematic cross section of an antiguided array. The antiguided cores correspond to the mesa (low-index) regions of the array [18].

In a negative-index guide, or antiguide, the high-order modes in the active region are stripped out by refraction, and the low-order modes are retained only because they have the least amount of refractive loss. In an array formed from antiguide devices, the optical field density in the areas between the devices (the "leaky modes") is high compared to that of the positive-index array, and lateral mode coupling is easiest to develop between devices in an antiguide array. Figure 23 [18] shows the negative-index guide array. The optical mode confinement in the antiguide is determined by the relative height of the refractive index barrier and the width of the antiguide. Radiation not confined is leaked into the area between antiguides, and strong mode coupling is obtained.

Along with the modal coupling within the high-index areas between antiguides is the possibility of parallel coupling, a resonance condition which allows leaky modes, and hence all the array elements, to equally couple to all others. This is accomplished by making the spacing between elements in the array equal to an integral half-wavelength of the leaky wave. When this spacing is met the area between elements becomes a Fabry-Perot resonator, being "pumped" by the lower-index antiguide region. This is the resonance optical waveguide (ROW) condition. The inter-element area becomes a resonator without having any reflections within this area. Since this condition mostly favors those modes which are at resonance, a high amount of modal discrimination is possible and the array elements operate at near in-phase.

Figure 24 [18] shows a schematic of a ROW laser. The devices are AlGaAs/GaAs based. A two-step growth process is used to prepare such a device. In the first step, separate confinement heterostructure–multi-quantum well (SC-MQW; see Section 7) array elements are prepared. Ridges are then chemically etched to define each element. The spacing between elements form what will be a passive waveguide in close proximity (within 0.2 μm) of the active region. This interelement area is fabricated from a separate process, and thus its refractive index can be controlled by adjusting the aluminum content of the alloy.

7 MULTIQUANTUM WELLS

In Section 2 it was observed that, in theory, the threshold current density drops as the active layer thickness is reduced until it reaches a minimum of ~0.1 μm. If the active layer thickness is reduced below this minimum value, the threshold increases because the dependence of optical confinement factor on active layer thickness. For a very small active layer thickness, separate means of confining photons and carriers are needed.

Decreasing the active layer width not only lowers the threshold but has been shown to provide other benefits, such as less temperature dependence, narrower gain spectrum, higher modulation frequency, and lower linewidth enhancement factor. This comes at the expense of increased complexity in diode design and difficulties in fabrication and processing of extremely thin films of semiconductors.

If several thin active layers are brought together, separated by thin barriers, a multi-quantum well structure is formed, as discussed in Section 14 of Chapter 1. In a single or multiple quantum well, the confinement of electrons into a two-dimensional space causes a quantization of the allowed energy levels and the formation of subbands of energy:

$$E_{l,n} = \frac{l^2 h^2}{8 m_\theta L_z^2} \tag{44}$$

where L_z is the thickness of each quantum well. The same expression holds for the valence band discontinuities. Recall the change in the density of states from parabolic to steplike given by

$$I_c(E) = \sum_{l=1}^{\infty} \frac{4\pi m_\theta}{h_z^2 L_z} H(E - E_{l,n}) \tag{45}$$

where $H(E - E_{l,n})$ is the Heaviside function. A plot of the density of states as a function of energy is shown in Fig. 25a for a bulk semiconductor and in Fig. 25b for a multi-quantum well. The density-of-states function narrows as a result of carrier quantization.

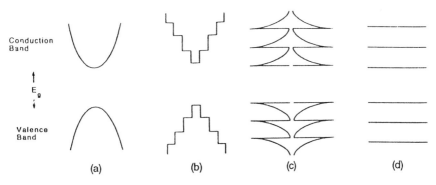

Figure 25 Examples of the ideal density-of-states functions for (a) bulk semi-conductor, (b) quantum well, (c) quantum wire, (d) quantum dot active regions. In each case, the increase of confinement of the carrier results in an increase in the singularity of the density of states.

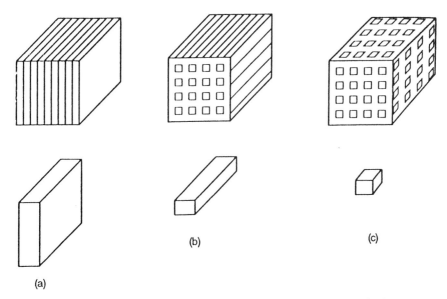

Figure 26 Examples of (a) quantum well, (b) quantum wire, and (c) quantum dot active regions of laser diodes. The structures are made by imbedding lower-band-gap material in a higher-band-gap matrix.

It is easiest to appreciate the effects of increasing the quantization of the electron and hole by comparing the different density of states for different quantization levels. Electrons and holes can be confined in one, two, or three dimensions, depending on the design of the heterostructure. This is illustrated for devices in Fig. 26. The quantum well, in Fig. 26a, is a structure where carriers are quantized in one dimension only. In the *quantum wire* and the *quantum dot* of Figs. 26b and c, the quantization is carried into the second and third dimentions, respectively. These struc-

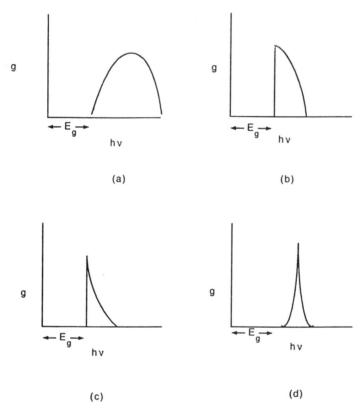

Figure 27 Spectral gain curves for lasers with (a) bulk semiconductor, (b) quantum well, (c) quantum wire, (d) quantum dot active regions. Increasing the singularity of the density of states in each case results in a narrowing of the gain curve and a lowering of the threshold current.

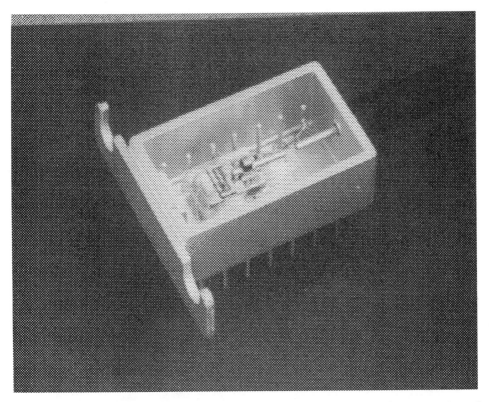

Figure 28 Commercial fiber-optic pigtailed laser diode. (Photo courtesy of Lasertron® Inc.)

tures are formed from lower-band-gap material imbedded in a higher-band-gap host. A quantum dot would confine the electron and hole to an area in three dimensions, of a size in each dimension on the order of the de Broglie wavelength.

The effect of increasing quantization in each dimension on the density of staes is shown in Figs. 25c and d. In each case, increasing the quantum confinement of the carrier in another dimension results in increasing the singularity of the density of states. For the quantum dot structure, the density of states (in theory) mimics that of a purely atomic system.

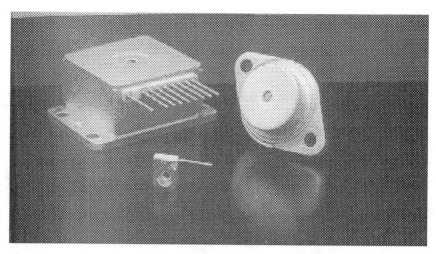

A

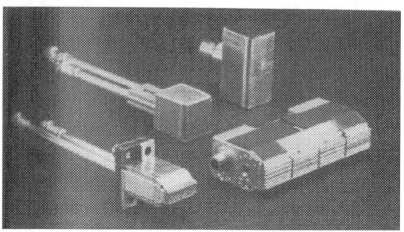

B

Figure 29 (A) 0.5- to 4-W laser diode in various heat sink, window, and fiber packages. (B) 50–500 W quantum well array laser diodes. (Photos courtesy of SDL® Inc.)

The results of narrowing of the density of states on laser performance is demonstrated in Eq. 13. The narrowing of the density of states results in a proportionally narrower gain spectra, as shown in Fig. 27. Gain narrowing is maximum for the quantum dot structure.

A narrow gain spectrum means fewer carriers are wasted in areas of the curve that do not contribute to the laser mode. Fewer wasted carriers in the active region lowers the value of n_{trans} necessary to achieve transparency, and the result is a significant lowering of the threshold current. Ultralow threshold lasers are then possible.

Increasing the quantization of the carrier using quantum wells, wires, or dots in the active region also increases the modulation bandwidth of the device. Recalling Eq. 35, the modulation corner frequency of a laser diode varies as the square of the parameter B, which is the differential gain dg/dn. The differential gain varies inversely with the width of the gain spectral curves of Fig. 27.

REFERENCES

1. S. Adachi, GaAs, AlAs, and $Al_xGa_{1-x}As$: material parameters for use in research and device applications, *J. Appl. Phys.*, *58*, R1 (1985).
2. H. Kressel and J. K. Butler, *Semiconductor Lasers and Heterojunction LEDs*, Academic, New York, 1977, p. 81.
3. F. Stern, Band tail model for optical absorption and the mobility edge in amorphous silicon, *Phys. Rev. B3*, 2636 (1971).
4. K. Vahala, L. C. Chiu, S. Margalit, and A. Yariv, On the linewidth enhancement factor α in semiconductor injection lasers, *Appl Phys Lett.*, *42*, 631 (1983).
5. N. K. Dutta and R. J. Nelson, The case for Auger recombination in $In_xGa_{1-x}As_yP_{1-y}$, *J. Appl. Phys.*, *53*, 74 (1982).
6. H. Kressel and D. E. Ackley, in *Device Physics*, Vol. 4 (C. Hilsum, ed.), North-Holland, Amsterdam, 1991, p. 725.
7. H. Kressel, Semiconductor lasers, in *Handbook on Semiconductors*, Vol. 4 (T. S. Moss and C. Hilsum, eds.), North-Holland, Amsterdam, 1981, p. 617.
8. F. Stern, Gain-current relation for GaAs lasers with n-type and undoped active layers. *IEEE J. Quantum Electron.*, *9*, 290 (1973).
9. J. E. Bowers and M. A. Pollack, Semiconductor lasers for telecommunications, in *Optical Fiber Telecommunications II* (S. E. Miller and I. P. Kaminow, eds.), Academic, New York, 1988, p. 509.
10. L. D. Westbrook, Dispersion of linewidth-broadening factor in 1.5 μm laser diodes, *Electron Lett.*, *21*, 1018 (1985).
11. K. Y. Lau, N. Bar-Chaim, I. Ury, Ch. Harder, and A. Yariv, Direct amplitude modulation of short cavity GaAs lasers up to X-band frequencies, *Appl. Phys. Lett.*, *43*, 1 (1983).

12. D. Botez, Laser-diodes are power-packed, *IEEE Spectrum*, 43 (June 1985).
13. H. Kressel, Semiconductor lasers, in *Handbook on Semiconductors*, Vol. 4 (T. S. Moss and C. Hilsum, eds.), North-Holland, Amsterdam, 1981, p. 636.
14. T. Yamada, T. Yuasu, M. Uchida, K. Asakawa, S. Sugata, N. Takado, K. Kamon, M. Shimizu, and M. Ishii, Fabrication and characteristics of dry-etched-cavity GaAs/AlGaAs MQW laser, in *Tenth IEEE International Semiconductor Laser Conference*, Kanazawa, Japan, paper PD4, 1986.
15. Y. Suematsu, K. Iga, and S. Arai, Advanced semiconductor lasers, *Proc. IEEE*, *80*, 383 (1992).
16. Bowers and Pollack, Semiconductor lasers for telecommunications, in *Optical Fiber Telecommunications II* (S. E. Miller and I. P. Kaminow, eds.), Academic, New York, 1988, p. 535.
17. N. Chinone and M. Nakamura, Mode stabilized semiconductor lasers for 0.7–0.8 and 1.1–1.6 μm regions, in *Semiconductors and Semimetals 22c* (W. T. Tsang, ed.), Academic, 1985, p. 61.
18. L. J. Mawst, D. Botez, P. Hayashida, M. Jansen, G. Peterson, T. J. Roth, J. Z. Wilcox, and J. J. Yang. Stabilized in-phase mode operation from monolithic antiguided diode laser arrays, in *Laser Diode Technology and Applications II*, Proceedings SPIE, vol. 1219 (D. Botez and L. Figueroa, eds.), Bellingham, WA, 1990, p. 127.

4

Photodetectors

1 INTRODUCTION

Most photodetectors can be classified as photon detectors or thermal detectors. In a photon detector, an absorbed photon excites an electron in such a way as to disturb the distribution of electronic charge in the detector medium. The detector works to measure the extent of this change in charge distribution, either by measuring a generated current or a generated voltage. Most solid-state photon detectors are still silicon based, due to low cost, flexibility in processing, and the excellent noise performance of silicon detectors.

In a thermal detector, the absorbed electromagnetic energy causes a change in temperature of the detector medium. In a thermistor bolometer, the temperature change is monitored through a change in detector resistance. In a pyroelectric detector, the change in temperature is detected by a change in the electrical polarization of the medium. Thermal devices have very low bandwidths and low quantum efficiencies. Since a blackbody has the ability to absorb a long range of wavelengths, they tend to have very broad spectral bandwidths, that is, a relatively flat response with respect to wavelength. Hence, they are most suited for applications requiring wide spectral bandwidths.

Photodetectors designed to operate in optical communication and information processing systems must possess high switching times and be sensitive to weak signals. The result has been the development of low-cost detectors which are only infrequently the limiting factor of the overall bandwidth of the communication system. For optical communications, the detector must be sensitive to a range of optical wavelengths between 0.7 and 0.9 μm or 1.3 to 1.55 μm. The latter range offers a one- to-two-order-of-magnitude decrease in silica fiber loss, but silicon is not sensitive beyond $\sim$1 μm and so is not usable as a detector material for this range. For this range the detector of choice is compound semiconductor based, normally as a *p-i-n* or avalanche detector.

There are three major categories of semiconductor-based photon detectors: photoconductors, photodiodes, and photon counters (photocathodes).

2 PHOTOCONDUCTORS

The basic principles of photoconductivity are illustrated in Fig. 1. Figure 1b illustrates the basic mechanisms of photogeneration that were discussed in Section 2 of Chapter 2. These generation processes represent

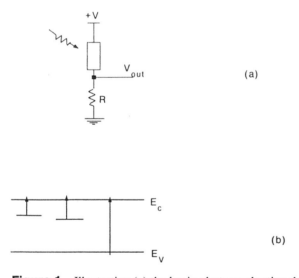

Figure 1 Illustrating (a) the basic photoconductive device in operation. A bias +*V* is applied to the device and a photocurrent generated by the incident photons is detected as an output voltage V_{out} across a fixed resistor *R*. (b) The basic photoexcitation mechanisms, including band-to-band and trap-to-band, responsible for generating the photocurrent.

the two different types of photoconductors: intrinsic and extrinsic. In each case, the absorbed electromagnetic radiation results in the promotion of a carrier to a higher energy state. With the change in state residing in the conduction band as illustrated in the figure, the result is a change in the conductivity, and thus a change in the resistance of the detector. The change in conductivity can be illustrated as

$$\Delta \sigma = \Delta n e \mu_n + \Delta p e \mu_p \tag{1}$$

where Δn and Δp are the change in the electron and hole free-carrier concentrations due to the presence of the optically generated carriers, and μ_n and μ_p are the electron and hole mobilities, respectively.

In an intrinsic photoconductor, the electromagnetic radiation is absorbed by a valence band electron, and the energy of the excitation is sufficient to promote the electron from the valence to the conduction band, creating an electron-hole pair. Since this promotion requires an incident photon energy E such that E is greater than the band-gap energy E_g, the sensitivity of the intrinsic photoconductor depends on the absorption coefficient, α, which is a strong function of wavelength. The spectral dependence of α is important in determining the most efficient photoconductive material for a given spectral range. Since the absorption coefficient for radiation of energy close to E_g is much higher for a direct- than for an indirect-gap material, preference is given to direct-gap compound semiconductors as the materials of choice for intrinsic photoconductors. One of the oldest and most successful optoelectronic devices has been the direct-gap CdS photodetector used for basic light sensing.

Intrinsic photoconductors may be used at room temperatures if the carrier lifetime is sufficient to overcome recombination and if the thermal noise associated with dark currents can be managed. As dark currents are larger in narrow-gap materials (especially for room-temperature operation of materials with $E_g < \sim 1$ eV), lower temperatures must be used for these detectors.

In an extrinsic photoconductor, the carriers are excited from bound impurity states to free conduction states. Examples of such detectors include germanium doped with elements such as Cu, Au, Hg, Zn, B, Ga or Sb. These detectors are used mainly for wavelengths beyond 5 μm, and suffer from problems of low sensitivity and low speed. They must be cooled to 77 K to minimize the thermal generation of carriers from the bound states.

In a free-carrier photoconductor, intraband excitation is made as illustrated in Fig. 2. The example of free-carrier absorption in this figure shows the doubly degenerate valence band commonly found in III-V compound semiconductors. If the valence bands E_k versus k close to the maximum

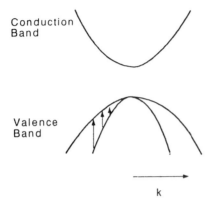

Conduction
Band

Valence
Band

k

Figure 2 Intraband absorption in the doubly degenerate valence band of a heavily doped *p*-type compound semiconductor.

of the band edge are taken as parabolic, one can approximate each band as

$$E_k = \frac{\hbar^2 k^2}{2m_k} \tag{2}$$

and thus the effective mass in each band is related to the band structure by

$$m_k = \frac{\hbar^2}{d^2E/dk^2} \tag{3}$$

The effective mass is different in each band having different curvature, d^2E/dk^2. Bands with higher curvature have carriers with smaller effective mass. The material shown in Fig. 2 is doped *p*-type, and transitions from the "heavy-hole" to the light-hole band are shown. A promotion of carriers from one band to another results in a change in the effective mass and a change in mobility, instead of a change in carrier density. The result is still a change in the conductivity, and an increase in the current which can be detected using the arrangement in Fig. 1a. Because of the narrow energy spacing in Fig. 2, photodetectors based on free-carrier absorption are capable of detecting radiation of energies well into the infrared.

A major drawback of the free-carrier photodetector is the necessity to minimize scattering by lattice vibrations. This necessitates operating the photoconductor below liquid nitrogen temperatures. Free-carrier photoconductivity has mainly been applied to InSb, which has a relatively large free-carrier absorption coefficient in the far infrared.

3 JUNCTION PHOTODIODES

Figure 3 illustrates the operation of a junction photodiode. Light is incident on the surface of a *p-n* junction, creating an electron-hole pair which is swept through the depletion region. Electron-hole pairs created by light close to, but outside the depletion region, may diffuse to the region. It is essential, though, that the majority of photons are absorbed within a carrier diffusion length of the depletion region of the junction. The carriers are swept through the junction producing either a short-circuit photocurrent or an open-circuit photovoltage, either of which is detectable.

In order for the incident radiation to be absorbed close to the junction, the junction must be close to the surface. The junction depth must be small compared to the carrier diffusion length, and indeed it is usually small compared to the optical absorption length. This can be accomplished by making the window material (usually the *p*-type side) very thin or by illuminating the device from the edge.

3.1 *p-i-n* Photodiode

A basic problem with the use of silicon as a photodetecting material is the indirect nature of its band gap. The indirect gap results in an absorption edge which is not sharp; i.e., the spectral response of the absorption coefficient does not cut off sharply at the band-gap energy. The spectral response of silicon is shown in Fig. 4 with the response of other direct-gap semiconductors. The result of a weak absorption edge is a smaller quantum efficiency for this wavelength as compared to direct-gap compound semiconductor photodetectors. To efficiently collect photons for which

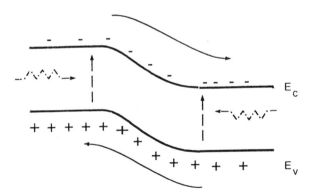

Figure 3 Basic photoexitation and drift of carriers in a *p-n* photodiode.

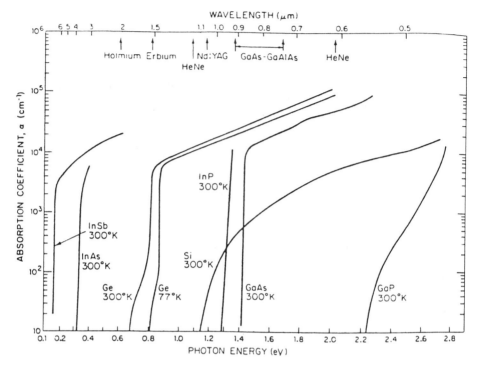

Figure 4 Spectral response of absorption coefficient for a variety of direct and indirect-band-gap semiconductors. (From Ref. 1.)

silicon has a low α, a long collection length must be included in the detector. This can be accomplished in a *p-n* photodiode by making the *n*-region long, as shown in Fig. 5a. A danger in making the *n*-side long is that the carrier diffusion length must be long enough for photoexcited carriers to reach the junction, or the junction depletion width must be widened to efficiently collect carriers. A low-doped *n*-type side will make the depletion width wide, or an intrinsic region can be inserted into the device. Either method provides a wide depletion region under no (or low) reverse bias. A *p-i-n* diode is prepared as seen in Fig. 5b. With the inclusion of the intrinsic region, the electric field is now spread over practically the entire device, instead of being localized at the *p-n* junction interface.

Details of photodiode characteristics will be covered in Sections 6 and 7, but parameters relevant to all photodetectors will first be reviewed.

(a)

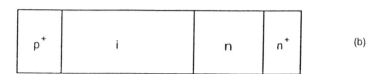

(b)

Figure 5   Structure of (a) a *p-n* and (b) a *p-i-n* photodiode.

4   BASIC PHOTODETECTOR CHARACTERISTICS

Most photodetection devices are classified by four attributes: sensitivity speed, overall efficiency, dynamic range. Sensitivity is determined by responsivity. Responsivity is measured as rms output current or rms output voltage per incident optical power, measured in amps/watt or volts/watt. For photoconductors and photodiodes operating in a photoconductive mode, an average photocurrent is measured and can be expressed as

$$I_p = \frac{q\eta_i P_p}{h\nu} \tag{4}$$

where P_p is the incident optical power and η_i is the internal quantum efficiency, or the probability that an incident photon results in the generation of a detectable free carrier. The responsivity of the device is

$$R = \frac{I_p}{P_p} = \frac{q\eta_i}{h\nu} \tag{5}$$

The efficiency η_i is mainly a function of the absorption coefficient and, hence, depends on the operating wavelength and depth of the absorption.

Other pertinent detector figures of merit besides the responsivity include the noise equivalent power (NEP), which is defined as

$$\text{NEP} = \frac{\text{noise}}{\text{responsivity}} = \frac{\text{radiant power}}{\text{signal/noise}}$$

or the optical power where the signal-to-noise equals unity. The NEP defines the minimum optical power the detector is capable of measuring. The specific detectivity $D*$ is also a measurement of minimal power capabilities, corrected for the area A of the detector and the bandwidth B:

$$D* = \frac{(AB)^{1/2}}{NEP} \tag{6}$$

It is a useful quantity for detectors where the noise power varies as $(AB)^{1/2}$. Another quantity often encountered is the specific noise equivalent power NEP*:

$$NEP* = \frac{1}{D*} \tag{7}$$

The speed or response time depends on the type of photodetector and the measurement technique in use. For photon detectors such as photo-conductors or photodiodes, transit time of the optically absorbed carrier across the device may be a limiting factor to the device response. However, as will be discussed, RC time constants and other time constants also exist.

The dynamic range of a photodetector is defined as the range between maximum and minimum optical inputs and is of most interest in digital applications. The minimum input is determined by the signal-to-noise ratio, while the maximum input is determined by the maximum quantity of light which can illuminate the detector without it being saturated. As speed of the detector may depend on optical intensity, there may be a trade-off between speed and dynamic range.

5   PHOTODETECTOR NOISE

The smallest detectable signal power in a photodetector is determined by the noise power. Figure 6 illustrates the various sources of noise possible in an optical detection system. Noise is present in the signal of interest and in the incident background light that is included with the signal. In the detector, noise sources are present and are amplified if there is a gain source internal to the detector, such as in an avalanche diode or photomultiplier. Noise is also present in the amplifier stages following the detector, as well as in any circuitry following the amplifier. It is desirable to reduce noise in the detector and amplifier so that the noise is limited by a single noise source, such as the statistical noise from the signal (so-called signal-limited noise) or the background of the signal (background-limited noise).

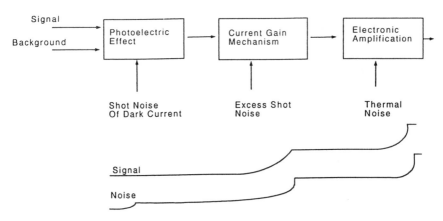

Figure 6 Phenomenistic model of noise in a photodetector.

There are various sources of noise from the photodetector itself. These may include, but are not limited to, shot noise, generation-recombination (g-r) noise, flicker ($1/f$), and normal Johnson noise. Shot noise may include any source of noise where the statistical fluctuations in the noise signal current obey a Poisson distribution. An example of shot noise is the statistical process of carriers being thermally or optically excited over a junction barrier, such as the barrier separating a *p-n* junction. Generation-recombination noise, also thermally or optically stimulated, is the noise associated with the creation or annihilation of electron-hole pairs across the band gap. These two sources of noise are impossible to separate, and constitute white noise, which has a flat power spectrum and is proportional to bandwidth and device area.

Another common noise source is flicker, or $1/f$ noise, where the noise spectrum has the $1/f$ dependence. This noise is associated with the trapping and re-emission of carriers to and from defect states at surfaces and interfaces at contacts, and is common to most all semiconductor devices. Flicker noise is process dependent and is reduced by maximizing processing conditions.

For a photodiode, Johnson noise is present in current flow through the neutral regions of the semiconductor away from the junction, but is also overshadowed by shot noise and g-r noise. Generation-recombination noise is present throughout the device, but is more prevalent in areas where the electric field is high. For example, it is more severe in the high *E*-field of avalanche diodes than in *p-i-n* photodiodes.

Detector noise is usually separated out into dark current noise—that is, noise present when illumination is off—and shot noise. Natural semiconductor activity dictates that at least two noise sources will be present influencing the dark current: noise due to random thermal variations in the carrier velocities, and noise from fluctuations in the g-r rates, but the latter g-r currents normally tend to dominate.

For a photoconductor detector, Johnson noise is present as it is for any normal conductor. The lack of a reversed-bias p-n junction in a photoconductor means that a substantial dark current may be present. There is a contribution from Johnson noise, therefore, associated with the finite conductivity of the photoconductor. Since the current noise limit of Johnson noise varies inversely with the resistance R of the conductor, the sensitivity of the photoconductor is then directly proportional to its conductivity and inversely proportional to the cross-sectional area. Maximizing these parameters to improve sensitivity unfortunately decreases frequency response.

6 PHOTODIODE OPERATING CHARACTERISTICS

Noise and responsivity ultimately determine operating bandwidth and detector characteristics. The responsivity of photodetectors is dependent upon the spectral response curves of Fig. 4. As mentioned earlier, to efficiently collect photons of low absorption coefficient requires a long p-n junction depletion width, which can be obtained by using a p-i-n junction. The junction can also be widened by the application of a reverse bias, and the field is increased by the presence of the intrinsic region within the junction. The intrinsic region ensures that the device is depleted across most of its bulk, and the reverse bias helps keep the electric field high.

There are two basic modes of photodiodes: the photovoltaic mode and the photoconductive mode. For the photovoltaic mode, the photoexcited electron-hole pair are drifted to opposite sides of the depletion region, where they increase the effective majority carrier concentration on each side of the junction. An open-circuit voltage generated by this buildup of charge is then measured. In photoconductive mode, a short-circuit current generated by the carriers is measured instead. Figure 7 shows the current-voltage properties of a photodiode under illumination by various photon flux. The I-V characteristics are given by the diode equation:

$$I_D = I_0 \left[\exp\left(\frac{qV}{kT}\right) - 1 \right] \tag{8}$$

where I_D is the diode current and I_o is the reverse saturation current.

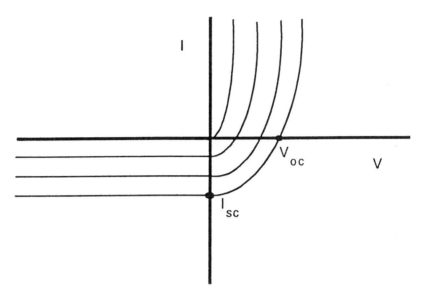

Figure 7 Current-voltage relationship for a photodiode under various photon flux. The diagram illustrates the open-circuit voltage V_{oc} and the short-circuit current I_{sc}.

If light generates electrons and holes at a rate G, those generated within a diffusion length of the junction will be collected and the short-circuit current I_{sc} can be determined as

$$I_{sc} = q(L_n + L_P)G \tag{9}$$

where L_n and L_p are the minority carrier diffusion lengths. In photovoltaic mode, if a finite resistance (such as the input resistance of an amplifier) is used to allow a photovoltage, V_p, to build up and be measured, a photovoltaic current I_p generated by the photovoltage can be measured as

$$I_p = I_{sc} - I_o \left[\exp\left(\frac{qV_p}{kT}\right) - 1 \right] \tag{10}$$

The open-circuit voltage can be determined as

$$V_{oc} = \frac{kT}{q} \ln\left[\frac{I_{sc}}{I_o} + 1 \right] \tag{11}$$

as shown in Fig. 7.

The responsivity, R, of a diode in photoconductive mode is given by

$$R = \frac{\eta_i q}{h\nu} \tag{12}$$

where η is the quantum efficiency of the detector. Unlike the photoconductive mode, in photovoltaic mode the responsivity depends on the operating point:

$$R = \frac{\eta_i q}{h\nu} r_d \tag{13}$$

where r_d is the dynamic resistance of the diode, given by the inverse slope of the I-V curve:

$$r_d = \left(\frac{\delta I}{\delta V}\right)^{-1}_{V = V_p} \tag{14}$$

Substituting the current-voltage characteristics for the photovoltaic device given by the diode equation yields a value of the dynamic resistance of

$$r_D = \frac{kT}{qI_o \exp(qV_p/kT)} \tag{15}$$

Since the responsivity in photovoltaic mode is V_p/P_p, where P_p is the average signal power, the measured signal voltage in photovoltaic mode will depend approximately on the logarithm of P_p.

The noise characteristics for a p-i-n diode can be described by the equivalent circuit model in Fig. 8, which is shown for a device in photoconductive mode. The input signal is detected by the device, which outputs a photocurrent I_p determined by

$$I_p = \frac{q\eta_i P_p}{h\nu} \tag{16}$$

where P_p is the average signal power from a modulated source modeled as

$$P_p(\omega) = P_p[1 + m \cos(\omega t)] \tag{17}$$

and m is the modulation index. Photocurrent generated by background radiation is I_B, and the dark current is I_D. The total statistical contribution to the shot-noise-generated noise current within a bandwidth B is then

$$\langle i_S^2 \rangle = 2q(I_p + I_B + I_D)B \tag{18}$$

If there is an amplified section within the diode (such as in an avalanche

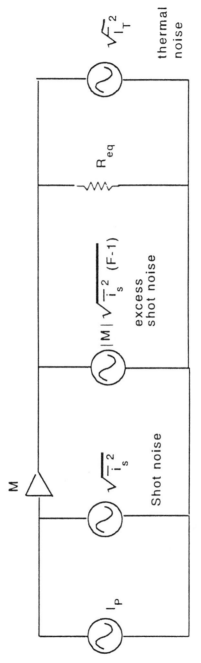

Figure 8 Equivalent circuit for determining the noise power of a photodetector.

device) there is an amplified noise

$$\langle i_n^2 \rangle = 2q(I_p + I_B + I_D)B|M(\omega)|^2 F(M) \tag{19}$$

where $|M(\omega)|$ is the ac power gain, and $F(M)$ is the excess noise factor which represents additional noise present in the amplified section. $F(M)$ is defined as

$$F(M) = \frac{S/N \text{ at output}}{S/N \text{ at input}} \tag{20}$$

where S/N is the signal-to-noise ratio. Strictly speaking, values of F and M are not the same for photo-, dark-, and background currents, due to their dependence on wavelength. Since one source will eventually dominate in the final result, we write them as equal for convenience. A final source of noise is the thermal noise of the equivalent resistance in the amplifier section:

$$\langle i_T^2 \rangle = \frac{4kTB}{R_{eq}} \tag{21}$$

where R_{eq} is an equivalent resistance which includes the resistance of the device and the input resistance of the following amplifier. A value for R_{eq} can be determined from the final measured power $P(\omega)$:

$$P(\omega) = \frac{1}{2} I_o(\omega)^2 R_{eq} \tag{22}$$

where $I_o(\omega)$ is the peak measured AC photocurrent.

Total noise current can be determined as $\langle i_n^2 \rangle + \langle i_T^2 \rangle$ and can be written in terms of the respective optical powers. The signal-to-noise ratio represented by the signal and noise optical powers is thus written [2]

$$S/N = \frac{\frac{1}{2}(q\eta_i P_p/h\nu)^2 |M(\omega)^2|}{[(2q^2\eta_i/h\nu)\,(P_p + P_B) + 2qI_D]|M(\omega)^2|FB + 4kTB/R_{eq}} \tag{23}$$

P_p is the signal optical power with $m = 1$ and P_B is the background optical power. From Eq. 23, a value for P_p can be determined in terms of S/N:

$$P_p = \frac{2B(S/N)h\nu F}{\eta_i}\left[1 + \left(1 + \frac{\phi_c}{B(S/N)F^2}\right)^{1/2}\right] \tag{24}$$

where

$$\phi_c = \left(\frac{\eta_i P_B}{h\nu} + \frac{I_D}{q}\right)F + \frac{2kT}{q^2 M^2 R_{eq}} \tag{25}$$

Here, ϕ_c represents all noise sources except the quantum noise from the signal itself. In the so-called signal limiting condition, the noise is limited only by the shot noise of the signal. These conditions exist when $\phi_c \ll B(S/N)F^2$, and in this limit the optical power is given by

$$P_p = \frac{h\nu B}{\eta_i} 4(S/N)F \tag{26}$$

The most fundamental limits of detection are set by the statistical fluctuations in radiation, both the signal and any background radiation. In this case the detector and amplifier noises are low. With advances in amplifier design, noise from the amplifier is low, and if the detector is operated under conditions where the background flux is less than the signal flux, then the fundamental limit to the noise is set by fluctuations in the signal. In practice the signal fluctuation limit is usually achievable only in cooled photomultiplier detectors with cooled amplifier stages. In the signal fluctuation limit, the noise equivalent power can be determined from Eq. 26 as

$$\text{NEP} = \frac{4h\nu BF}{\eta_i} \tag{27}$$

The above expressions for radiation-limited performance are valid in a junction diode in reverse bias only when amplifier noise is neglected and the dark reverse saturation current is much less than the illuminated reverse saturation current. For a p^+-n junction the latter can be expressed as

$$\eta_i \phi_B \gg \frac{nW}{\tau_p} \tag{28}$$

where W is the depletion width, τ_p is the minority carrier lifetime, n is the electron concentration in the n-side, and ϕ_B is the background photon flux. Radiation-limited noise is not possible in an avalanche photodiode as the current gain process provides noise in excess of the signal noise. The criterion for radiation-limited performance in a photoconductor is complicated by the noise contribution from recombination-rate statistics.

One way to obtain radiation-noise-limited performance is to cool the detector to low temperatures, where dark currents are small, and operate under conditions of low background flux. Material with a low concentration of recombination centers is required to minimize dark currents. Higher-temperature operation can only be obtained if the efficiency of the device is very high and the doping is relatively light.

6.1 Dark Current/Background/Johnson-Noise-Limited Performance

A much more common condition is that the noise is limited by shot noise from the background or from Johnson noise rather than from the signal quantum noise. In this condition, ϕ_c of Eq. 25 is such that $\phi_c \gg B(S/N)F^2$ and the average optical power level is given by

$$P_p = \frac{2h\nu}{\eta_i} (B(S/N)\phi_c)^{1/2} \tag{29}$$

Under certain conditions, the signal may be buried in a large amount of background noise. If the other noise sources are minimized, the noise is then dominated by the background photon flux. The so-called background-limited IR photodetector (BLIP) has a noise equivalent power which can be expressed as

$$\mathrm{NEP}_{\mathrm{BLIP}} = \left(\frac{2h\nu BFP_B}{\eta_i}\right)^{1/2} \tag{30}$$

Equation 30 is appropriate for a detector operating at a background temperature T and a signal energy such that $h\nu \gg kT$ (i.e., ~ 50 μm or less for a 300 K background). BLIP performance is the most fundamental of photodetection and is often the most desirable to obtain for detectors used in mid- and far-infrared regions of the spectrum where the signal represents a warm entity which must be detected against a thermal background. Reaching this limit in the mid- and far-IR requires cooling the detector and amplifier, limiting the detector bandwidth, and placing a cooled shroud around the detector opening.

It has been shown [3] that a photodiode of internal capacitance C operating under wideband conditions has a value of internal equivalent resistance R_{eq}, which is limited by (ignoring the amplifier noise)

$$R_{\mathrm{eq}} \leq \frac{1}{4BC} \tag{31}$$

Because of this low value of R_{eq}, a photodiode operating under wideband conditions is always thermal noise limited. If the current gain of the detector is high enough, the contributions from other noise sources (background-limited noise and dark current noise) can be increased until it is on the order of the Johnson noise term. Increase of the current gain above this point is undesirable as the added noise of the multiplication increases, and the bandwidth becomes limited at the higher gain. The NEP is then roughly twice the value in Eq 30. This sets a maximum limit on internal current gain.

If the noise is dominated by thermal noise, then the detector is in the Johnson-noise-limited (JOLI) regime, and the NEP is given as

$$\text{NEP}_{\text{JOLI}} = \frac{2h\nu(kT)^{1/2}B^{1/2}}{\eta_L g R_{\text{eq}}^{1/2} M} \tag{32}$$

It is often more convenient as well as educational to express the NEP and D^* in terms of the optical flux instead of optical power. Using the relation

$$\phi_B = \frac{P_B}{h\nu A} \tag{33}$$

for a detector of area A and a flux of ϕ_B, values of NEP and D^* have been summarized in Table 1.

For fiber-optic detectors, or other applications where background radiation is effectively filtered, and which operate at wavelengths where warm background is not a limiting factor, the main contribution to shot noise will be from the dark current of the diode. Sources of dark current include crystal defects and imperfections, impurities in the depletion re-

Table 1 Values of NEP and D^* in Terms of the Optical Flux for Signal Limiting, Background Limiting (BLIP), Johnson-Noise-Limited (JOLI), and Amplifier-Limited Conditions

Operating conditions	NEP	D^*	characteristics
Signal limiting	$\dfrac{4Fh\nu B}{\eta}$	$\dfrac{\eta(A/B)^{1/2}}{4Fh\nu}$	Specialized low-bandwidth, low-temperature detectors.
Background limited	$h\nu\left(\dfrac{2A\phi_B B}{\eta}\right)^{1/2}$	$\dfrac{1}{h\nu}\left(\dfrac{\eta}{2\phi_B}\right)^{1/2}$	Where bandwidths are low and background flux is high, or for APDs with high internal gain.
Johnson noise limited	$\dfrac{2h\nu(kT)^{1/2}B^{1/2}}{\eta q R_{\text{eq}}^{1/2} M}$	$\dfrac{\eta q R_{\text{eq}}^{1/2} A^{1/2} M}{2h\nu(kT)^{1/2}}$	Usual operating conditions for p-i-n diodes.
Amplifier limited	$\dfrac{4\pi h\nu C_T(I_3 kT)^{1/2}B^{3/2}}{\eta q g_m^{1/2} M}$	$\dfrac{\eta q g_m^{1/2} A^{1/2} M}{4h\nu(kT)^{1/2}I_3^{1/2}BC_T}$	Prominent for high bit rates. Amplifier noise is coupled with diode capacitance

gion, and diffusion of carriers out of the high-resistivity sidewalls of the depletion region. By use of controlled processing, dark currents are usually minimized. However, since the dark current is thermally activated, temperature is a key issue in reducing noise. Figure 9 shows a plot of dark current versus temperature for an avalanche photodiode. Difference of the slopes of the curve represent different thermally activated processes contributing to the current. For a typical high operating temperature for photodiodes (between 40 and 150°C), the dark current for this particular device is dependent on carrier diffusion. For operation at lower temperatures, the dark current is determined by traps with energy levels close to the middle of the band gap. Figure 10 shows noise equivalent photodiode dark current versus bit rate. The raise in the curve above 100 Mbit/s represents the increased noise contribution due to the amplifier resistance.

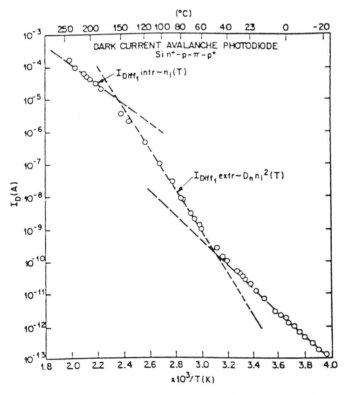

Figure 9 A plot of dark current as a function of reciprocal temperature illustrating the various thermally activated processes contributing to the dark current of an avalanche photodiode. (From Ref. 4.) © 1978 AT&T. Reprinted with permission of AT&T.

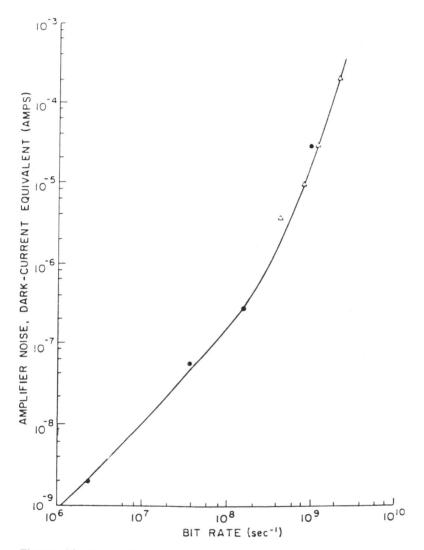

Figure 10 Upper limit on the acceptable photodiode dark current for various bit rates when the amplifier noise dominates the system sensitivity. (From Ref. 5.)

For a device which has internal current gain, such as an avalanche photodiode, the dark current noise contribution will be a substantial part of the detector noise, being almost equal to the Johnson noise. The main emphasis, then, is to reduce the total diode noise so it is on the order of the amplifier noise.

We have so far ignored amplifier noise. Forrest [6] has derived a more detailed expression for the noise associated with a *p-i-n* diode followed by a preamplifier which has an FET input, a typical combination for high-bandwidth operation. In this analysis, the shot noise in Eq. 18 has included the leakage current of the FET front end and the total noise includes thermally generated currents from the FET channel. In this analysis the noise of the preamplifier-diode combination at high bit rates B can be written as

$$\langle i_c \rangle^2 = \frac{4kT\Gamma}{g_m}(2\pi C_T)^2 I_3 B^3 \tag{34}$$

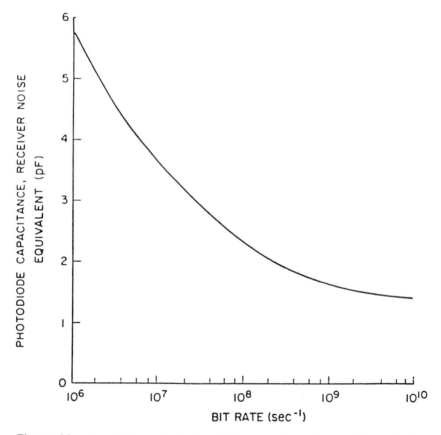

Figure 11 Acceptable equivalent amplifier capacitance for amplifier noise dominated system sensitivity vs. bit rate. (From Ref. 5.)

where g_m is the transconductance of the FET, Γ is the excess noise factor of the FET, I_3 is a Personick integral, and C_T is the total capacitance of the amplifier and photodiode. The strong dependence of Eq. 34 on B will mean this noise term dominates at extreme high bit rates. Pearsall and Pollack [5] have analyzed the maximum allowable capacitance of a photodiode which would be required to keep the capacitance noise term of Eq. 34 below the Johnson noise term alone, and plotted this capacitance as a function of bit rate in Fig. 11. For bit rates greater than 200 Mbit/s, the lowest allowable photodiode capacitance would be ~0.2 pF, and an increase in capacitance to 0.5 pF would represent a loss of ~1 dB. In terms of a long-distance fiber-optic link, this represents a decrease of the longest allowable distance between repeaters of ~5 km.

To summarize, for most photodiode applications, BLIP performance will only be obtained if the internal gain of the photodiode is enough to overcome the Johnson noise. JOLI performance will dominate unless the total diode noise is less than the amplifier noise. In the latter case, particularly dominant at high bandwidths, the system is limited by the amplifier plus photodiode capacitance.

7 AVALANCHE PHOTODIODES (APD)

The avalanche photodiode is a unique solid-state device, as it is one of the few photodetectors, and the only photodiode, with internal current gain. The diode operates at high reverse bias with a large internal electric field. The gain is provided by an internal-gain-producing mechanism termed *impact avalanche* process.

As discussed in Section 6, for a traditional high-bandwidth photodiode operating at high frequencies, the overall noise is limited by the Johnson noise associated with the resistance of the biasing resistor and the noise of the following amplifier stage. For an APD, the internal gain of the diode can be increased until the shot noise produced equals the Johnson noise, and the signal power can be increased without an appreciable increase in the noise power as compared to photodiodes with no internal gain. The ideal quantum noise limit represented by Eq. 27 cannot be obtained because of the excess noise surrounding the avalanche multiplication process itself. For silicon APD devices, values of the internal gain may be as high as 100, and this increases the gain-bandwidth product above those devices not containing the gain mechanism. Today's silicon avalanche photodiodes have gain-bandwidth products as high as 100 GHz. Figure 12 shows a comparison of various devices used in fiber-optic applications. The application where an APD is used as a receiver results in a longer distance between repeaters used to reamplify a signal along a long fiber distance.

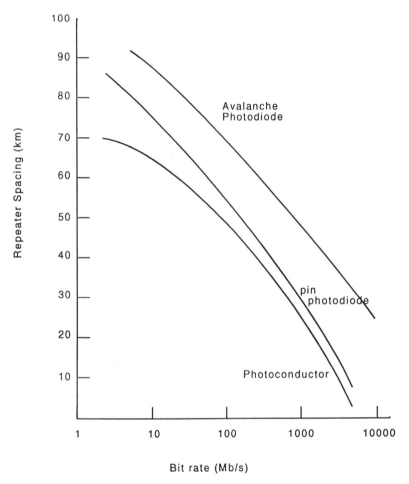

Figure 12   Allowable repeater spacing for a long-haul fiber-optic system vs. bit rate for different photodetector types. (After Ref. 7.)

The impact avalanche gain mechanism is shown schematically in Fig. 13. In Fig. 13a, an electron entering from the p-side is accelerated to a high kinetic energy from the large internal electric field. The electron gains enough energy to ionize an atom upon collision with the lattice. This ionizing collision has sufficient energy that an electron is promoted from the valence to the conduction band and an electron-hole pair is created. The electrons are swept toward the n-type side and the hole toward the p-side. This single collision is called impact multiplication. If, however,

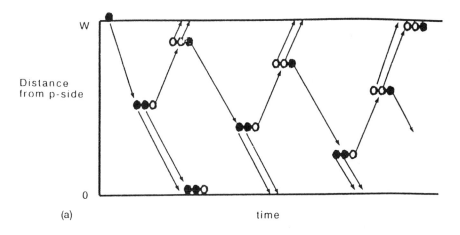

(a)

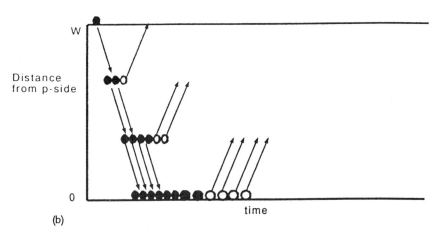

(b)

Figure 13 Schematic of the impact-avalanche process across the depletion region. Distance is measured from the edge of the neutral p-side. In the representation (a), both carriers contribute equally to the ionization process ($\alpha = \beta$). In (b) only the electron contributes to impact ionization ($\beta = 0$).

the generated carriers gain enough energy to cause further impact-ionizing collisions with the lattice, the result is further creation of electron-hole pairs. Each generated electron and generated hole can create a new electron-hole pair, each of which has the ability to create another pair, etc. This process is called avalanche multiplication.

In Figure 13a both the electron and hole are represented as having an equal probability of ionizing an atom. In practice one carrier (such as the electron) may have a higher probability of an ionization collision than the other. Figure 13b shows a schematic where the probability of a hole having an ionizing collision is zero. The probability that an electron and a hole traveling a distance dx measured along the direction of the electric field has enough energy for an ionizing collision is $\alpha\,dx$ and $\beta\,dx$, respectfully, where α and β are the ionization coefficients, or ionization rates, of the two carriers. The values of α and β, along with the interacting length of the collision process and the carrier density, determine the amount of avalanche gain. Measured values of α and β for silicon and germanium are shown in Fig. 14.

For the process in Fig. 13a, $\alpha = \beta$ and this is termed a "symmetric" ionization process. As the impact multiplication process is probabilistic, there are random fluctuations in the distance between successive collisions, and the multiplication factor for both electrons and holes is widely distributed. This represents a finite distribution in the number of carriers generated, and hence a noise current. This noise is lowest in materials that are asymmetric (Fig. 13b), since if one carrier dominates the ionization process the resulting multiplication factor distribution is narrower. Thus, asymmetric materials are preferred for avalanche devices.

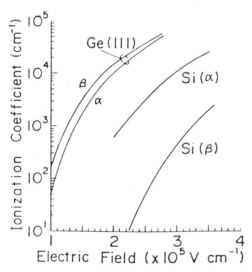

Figure 14   Impact ionization coefficients for silicon and germanium at 300 K. (From Ref. 8.)

As seen from Fig. 14, the probability of ionizing collisions in silicon is 40–50 times higher for electrons than for holes. The relative ionization energies are less symmetric for materials with indirect band gaps; thus most compound semiconductors are less suitable for use as APDs. Furthermore, the band-to-band tunneling probability is higher in direct-gap semiconductors, and the avalanche process must compete against this alternative breakdown mechanism under the high electric fields necessary for impact ionization. Germanium, also with more symmetrical ionization rates, has more noise as an APD than silicon. As Ge is used as an APD device for 1.0–1.6 micron applications, it must compete with the excellent GaInAs *p-i-n* diode as a detector for that wavelength.

The minimum energy needed for impact ionization is called the ionization threshold energy. In order to produce an ionizing collision, the carrier must have a threshold energy which is about 1.5 to 1.8 times the value of the band-gap energy. Since the probability of a scattering collision is far greater than that of a ionizing collision, impact ionization must compete against scattering processes. As the temperature increases, the amount of scattering from phonons increases, values for the ionization coefficients then decrease, and then the avalanche gain decreases.

The electric field required to achieve impact ionization is dependent upon the band gap of the semiconductor and may range from 10^4 to 10^5 V/cm for photodiodes using avalanche gain.

Wolff [11] has solved the Boltzmann equation for the impact-ionizing processes and determined a relation for the ionizing coefficients. Wolff assumed two basic scattering processes for carriers: impact ionization and scattering by optical phonons. Wolff determined a relationship

$$\alpha(E) = A \, \exp\!\left(\frac{-b}{E}\right)$$

for the coefficient, with E as the electric field. The constants A and b must be found by experiment. The parameters α and β are exponential functions of the electric field, and hence, depending on the structure, the internal current gain is not a linear function of the applied voltage. Lee et al. [12] determined measured values for α and β for silicon APDs:

$$\alpha = 3.8 \times 10^6 \exp\!\left(\frac{-1.75 \times 10^6}{E}\right) \text{cm}^{-1} \tag{35}$$

and

$$\beta = 2.25 \times 10^7 \exp\!\left(\frac{-3.26 \times 10^6}{E}\right) \text{cm}^{-1} \tag{36}$$

Avalanche photodiodes can be characterized by three basic properties: quantum efficiency, response speed, and noise. Quantum efficiencies of APDs are similar to photodiodes with no internal gain and are thus characterized by the parameter η_i and depend on values of the absorption coefficient for the detecting wavelength. Response speeds are dependent on the RC time constant of the diode and amplifier circuits, transit times of the carriers in the depletion region, and carrier diffusion times, but they also depend on the buildup time for the avalanche process. Noise is governed by multiplication noise.

7.1 Multiplication

The multiplication process of APDs can be modeled using classical continuity analysis. Electrons and holes, in the physical representation of Fig. 13, are photogenerated outside a high-field region and are injected into this region, which has width W. Electrons are injected from the p-side at $x = 0$ and travel in the positive x direction, and holes injected from $x = W$ travel in the negative x direction. In Fig. 13a, the change in the electron current with distance for electrons generated by both carriers in the depletion region can be modeled, for uniform electric fields, as

$$\frac{dJ_n(x)}{dx} = \alpha J_n + \beta(J - J_n) \tag{37}$$

where the total current J is the sum of the electron and hole currents, J_n and J_p, respectively. Equation 37 is a differential equation in J_n which has solution

$$J_n(x) = \frac{J \int_0^x \beta \exp\left[-\int_0^x (\alpha - \beta)\, dx'\right] dx + C}{\exp\left[-\int_0^x (\alpha - \beta)\, dx'\right]} \tag{38}$$

where C is a constant of integration. Using the boundary conditions $J_n(W) = J_p(0) = J$, and $J_n(0) = C$, and with the identity

$$\exp\left[-\int_0^W (\alpha - \beta)\, dx'\right] = 1 - \int_0^W (\alpha - \beta) \exp\left[-\int_0^W (\alpha - \beta)\, dx'\right] dx \tag{39}$$

we can solve for the ratio $J/J_n(0)$, which is equal to the multiplication factor for electrons M_n:

$$M_n = \frac{J}{J_n(0)} = \left[1 - \int_0^W \alpha \exp\left[-\int_0^W (\alpha - \beta)\, dx'\right] dx\right]^{-1} \tag{40}$$

A similar expression for the multiplication factor for holes can be derived:

$$M_p = \left[1 - \int_0^W \beta \exp\left[- \int_0^W (\alpha - \beta)\, dx'\right] dx\right]^{-1} \tag{41}$$

An important condition for APD operation is control of the avalanche multiplication effect. The threshold voltage for avalanche gain operation must be controlled, and the tendency for avalanche breakdown must be avoided. Avalanche breakdown is defined as the condition that M_n or $M_p \rightarrow \infty$, or where the denominator of Eq. 40 (or Eq. 41) approaches 0.

The usual method of illustrating the effects of α and β on device design is to first assume $\beta = 0$ (or $\alpha = 0$); that is, one carrier completely dominates the ionization process. Under the conditions that only injected electrons are involved in impact ionization, Eq. 40 reduces to

$$M_n = \exp\left[\int_0^W \alpha\, dx\right] = \exp(\alpha W) \tag{42}$$

Under these conditions, there is no possibility of avalanche breakdown as there is always a limit to the maximum value of M_n. The avalanche gain increases exponentially with depletion width W. The response time of the device is dependent on the transit time of the carrier, which is the time it takes the first injected electron and the last ionized hole to cross the depletion region, i.e., the sum of the electron-hole transit times or twice what it takes for a photodiode with no gain. Since each electron must create twice as many carriers to produce the same value of gain as a device where $\alpha = \beta$, more carriers are present and the distance between successive ionizing collisions is less. The distribution of these collision distances is narrower, and the noise from the avalanche process is then reduced. Since the gain is independent of the transit time of the carriers there is no gain-bandwidth product and no positive feedback for the case $\beta = 0$ (or $\alpha = 0$).

For a material where $\alpha = \beta$, both M_n and M_p must be considered and each reduce to

$$M_n = M_p = \frac{1}{1 - \int_0^W \alpha\, dx} = (1 - \alpha W)^{-1} \tag{43}$$

and avalanche breakdown is obtained for $W\alpha = 1$. On average, each electron injected into the high-field region creates an electron-hole pair, and each generated carrier creates another pair. This represents a positive-feedback condition, as high-energy holes are constantly being regenerated, and thus is an unstable system. Under breakdown, the gain is infinitely high and the completion time for the injected carrier to cross the high-field region is infinitely long, so the bandwidth is infinitely narrow. This dependency of gain on transit time forces a finite gain-bandwidth product. The gain-bandwidth product then becomes roughly equal to the reciprocal of the average of the two carrier transit times [13].

In a practical system, the exact process will be somewhere between these two extremes. The device designer must ensure that the undesirable conditions of positive feedback are minimized so breakdown does not occur. In the case where α and β are both finite but are assumed constant across the high-field region, Eq. 40 can be written as

$$M_n = \frac{[1 - (\beta/\alpha)]\exp\{\alpha W[1 - \beta/\alpha]\}}{1 - (\beta/\alpha)\exp\{\alpha W[1 - \beta/\alpha]\}} \tag{44}$$

and the positive-feedback factor becomes β/α. Because the electric field under normal operation of the device is high enough that the velocity of both carriers is saturated and so independent of position [13], constant ionization coefficients are a valid assumption even though the electric field is exponential in the high-field region.

Webb et al. [9] have made a slightly more useful form of Eq. 40 by writing it as

$$M_n = \frac{\exp[-(1 - k_0)\delta](1 - k_1)}{\exp[-(1 - k_0)\delta] - k_1} \tag{45}$$

where

$$k_0 = \frac{\displaystyle\int_0^W \beta \, dx}{\displaystyle\int_0^W \alpha \, dx} \tag{46}$$

$$k_1 = \frac{\displaystyle\int_0^W \beta \exp\left[-\int_0^W (\alpha - \beta) \, dx'\right] dx}{\displaystyle\int_0^W \alpha \exp\left[-\int_0^W (\alpha - \beta) \, dx'\right] dx} \tag{47}$$

and

$$\delta = \int_0^W \alpha \, dx \tag{48}$$

Thus, k_0 is an average and k_1 is a weighted average of the hole and electron ionization rates. Fortunately, k_0 and k_1 can be considered constant and equal. Thus, values for the multiplication factor can be determined.

The electron multiplication factor M_n calculated from Eq. 46 for a typical diode model is shown in Fig. 15. The reverse-bias voltage in the model varies from 20 to 45 V across a 1-μm-wide intrinsic region. Because of the dependence of α on the electric field, the curve has been plotted for various ratios of α/β. M_n is an exponential function of the electric field, and the slope of the curve is a strong function of k_1.

Webb also pointed out the difference that small fluctuations in processing conditions can have on the multiplication factor. For $\beta/\alpha = 0.01$, at a reverse bias of 100 V, a 0.5% variation in the electric field would result in a 20% variation in multiplication. For $k_1 = 1$, the variation in multiplication is raised to 320%.

Figure 16 shows the dependence of M on the position in the high-field region for each of the extremes in ionization coefficients. For equal

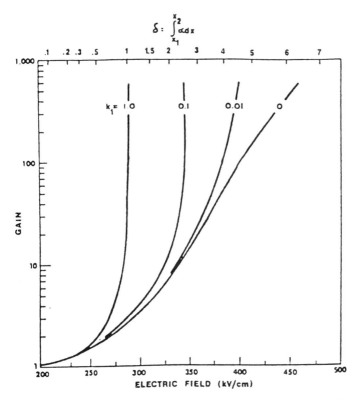

Figure 15 Electron multiplication factor M as a function of the electric field for various ionization ratios. (From Ref. 9.)

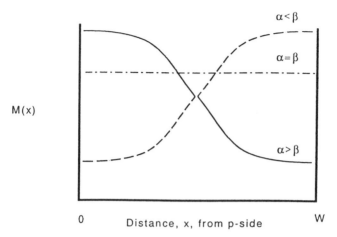

Figure 16 Multiplication factor as a function of position across the high-field region, for various ionization ratios.

ionization coefficients, the multiplication is constant across the region. For unequal coefficients, it is desirable to generate the majority of electrons (in the case $\alpha > \beta$) close to the edge of the p-region in order to maximize the gain. Where the electrons are generated depends on the device design—that is, the distance of the p-edge from the surface of the device and hence on the value of the absorption coefficient. Thus there is also a dependence of the multiplication factor on the wavelength of the incident radiation.

7.2 Frequency Response

Response times for avalanche diodes depend on the same three parameters as diodes with no internal avalanche gain: (1) diffusion time of carriers generated outside the depletion region; (2) transit time of carriers across the depletion (or high-field) region; (3) internal RC time constant of the device. All three of these parameters depend on the width of the depletion region and, hence, as pointed out in Section 3, on the absorption coefficient of the incident light. For silicon, the length of the depletion region may be 30–50 μm, and hence response times are dependent on transit time of the carriers alone.

The fourth parameter governing response time of APDs is the buildup time of the multiplication process. This time also depends heavily on the ratio of the ionization coefficients. The multiplication buildup time, t_m, is related to the multiplication factor by

$$t_m = \tau \langle M \rangle \tag{49}$$

where $\langle M \rangle$ is the average multiplication factor determined by

$$\langle M \rangle = \frac{J_p M_p + J_n M_n}{J} \tag{50}$$

Emmons [13] has shown that the intrinsic response time τ can be expressed as

$$\tau = N \left(\frac{\beta}{\alpha} \right) \left(\frac{W}{v_s} \right) \tag{51}$$

where v_s is the carrier saturation velocity, W is the depletion width, and N is a number which varies from 0.33 to 2 as the ratio β/α varies from 1 to 10^{-3}.

Transit times for carriers can be determined by solving the time dependence of the continuity equation represented by Eq. 37. The frequency variation of multiplication considering transit time effects alone is given by

$$M(\omega) = \frac{M_n}{(1 + \omega^2 M_n^2 \tau^2)^{1/2}} \tag{52}$$

Emmons has calculated values of $M(\omega)$ for various values of α/β and the results are shown in Fig. 17. Here the intrinsic parameter τ was taken as the average of the electron and hole transit times t_n and t_p. In turn, these parameters were determined by assuming the saturation velocities were the same as both carriers and taken as equal to W/v_s. For values of M_n in Fig. 17 such that $M_n < \alpha/\beta$, the bandwidth is nearly independent of gain and the response is dependent solely upon the transit time of the carriers. For values of $M_n > \alpha/\beta$, the gain bandwidth product at high frequencies is given by

$$M_n(\omega)B = \frac{M_n}{N(W/v_s)(\beta/\alpha)} \tag{53}$$

and thus the response is limited by avalanche buildup effects. To obtain high gain-bandwidth products under these conditions, v_s should be large and β/α and W should be small.

As in unmultiplied *p-i-n* structures, effects of the diffusion of carriers on response time can be minimized by keeping the device under full depletion, thus minimizing the length of the undepleted region of the device. This is an essential requirement for high-speed operation, as diffusion

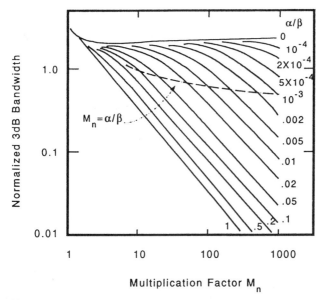

Figure 17   Three-decibel bandwidth as a function of the multiplication factor for various ionization ratios. (After Ref. 10.)

times of carriers in undepleted material are on the order of the reciprocal of the minority carrier lifetimes.

Similar to nongain photodiodes, the RC time constant can be evaluated as

$$t_{RC} = (R_s + R_L)C \tag{54}$$

where R_s is the diode series resistance and R_L is the load resistance. The value of C includes the diode junction capacitance and the connector capacitance, the former of which should be the governing value for determining values for C. Since the junction capacitance is inversely proportional to W, the depletion layer should be made long for small values of t_{RC}. As increasing W increases the transit time of carriers, as well as influencing the gain, the trade-off concerns an optimum value for W. Fortunately, the depletion-layer capacitance can also be controlled by control of the doping in the device.

The actual frequency response, then, is limited by the operating wavelength (the value of the absorption coefficient for this wavelength determines the value of the depletion width), the low-frequency avalanche gain, and the semiconductor device parameters and device geometries which affect t_{RC}.

7.3 Excess Noise Factor *F*

As mentioned earlier, the amount of noise expected in an APD depends on the probability distribution of collision events, and hence on the β/α ratio. Since not every generated pair experiences the same collision, and hence does not give the same contribution to the value of the gain, the mean square multiplication factor $\langle M^2 \rangle$ becomes greater than the average of the multiplication squared, $\langle M \rangle^2$. Since the noise is dependent on the mean square multiplication factor, a parameter to measure the contribution to noise from the multiplication process is the excess noise factor *F* defined as

$$F = \frac{\langle M^2 \rangle}{\langle M \rangle^2} \tag{55}$$

F has a value of 1 if there is no gain process or if each collision exactly results in a gain of M_n or M_p.

The excess noise factor has been examined by McIntyre [14]. For an incremental increase in electron and hole currents dI_n and dI_p in a distance dx, the probability that a carrier will undergo an ionizing collision and create an electron hole pair is governed by Poisson statistics. Thus the current generated in dx produces a shot noise $2qBdI_n$ within a bandwidth *B*. The resulting change in the noise spectral density, $d\phi$, is given by

$$d\phi(x) = 2qM^2(x)\, dI_n(x) \tag{56}$$

By integrating the continuity equation given by Eq. 37, a value for *F* can be obtained as

$$F = \frac{\phi}{2qI_0 \langle M \rangle^2} \tag{57}$$

where I_0 is the total current and the contribution to the noise spectral density from multiplication noise is given by

$$\phi = 2q \left[2I_n M^2(0) + 2I_p M^2(W) + 2 \int_0^W G(x) M^2(x)\, dx \right.$$

$$\left. + I_0 \left\{ 2 \int_0^W \beta M^2(x)\, dx - M^2(W) \right\} \right] \tag{58}$$

Meaningful data was obtained by McIntyre by again rewriting the equation in terms of weighted averages which are constant across the depletion region. That is, defining new parameters k_1 and k_2 such that

$$k_1 = \frac{\displaystyle\int_0^W \beta M(x)\, dx}{\displaystyle\int_0^W \alpha M(x)\, dx} \tag{59}$$

$$k_2 = \frac{\displaystyle\int_0^W \beta M^2(x)\, dx}{\displaystyle\int_0^W \alpha M^2(x)\, dx} \tag{60}$$

the excess noise factor can then be evaluated for injected electrons and holes. The factor for electrons, F_n, is determined for electrons injected into the high-field region from the p-side, and the hole noise factor F_p for holes injected from the n-side as

$$F_n = kM_n + (2 - M_n^{-1})(1 - k) \tag{61}$$

$$F_p = k'M_p + (2 - M_p^{-1})(k' - 1)$$

where

$$k = \frac{k_2 - k_1^2}{1 - k_2} \cong k_2 \tag{62}$$

and

$$k' = \frac{k}{k_1^2} \tag{63}$$

F_n is plotted as a function of M_n for different values of k in Fig. 18. Since k depends roughly on β/α, if low noise is desired, only the carrier with the highest ionization coefficient should be injected into the high-field region.

To summarize, there are several reasons why it is desirable for one carrier to have a greater ionizing coefficient than the other. These include the ability to maintain control of the threshold voltage without going into breakdown, the ability to minimize response times, and the ability to minimize noise figures. The device should be designed so that only the carrier with the highest ionization coefficient is injected in the high-field region, and this initial carrier must be injected into the region where the field (and hence the multiplication) is at a maximum. In addition, to ensure maximum efficiency, the absorption coefficient at this wavelength should be high. Unfortunately, few semiconductors meet all these desirable qualities.

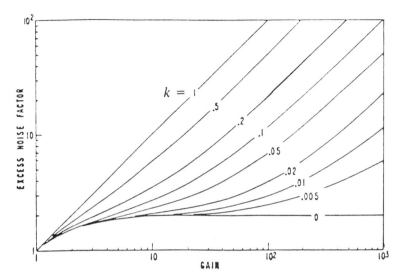

Figure 18 Excess noise factor F as a function of the electron gain M_n for various ionization ratios. (From Ref. 9.)

7.4 APD Structures

Avalanche devices have been fabricated in a variety of structures. Among these are p-n junctions, Schottky barriers, MOS structures, p-i-n diodes, p^+-π-p-n^+, n^+-π-n-π-p^+ structures (π-type signifies lightly doped p-type).

Because of the nature of the absorption coefficient for silicon, a large-volume absorption region is necessary to obtain a high quantum efficiency. For detecting 0.8–0.9 μm radiation, a depletion width of 30–50 μm is necessary for a quantum efficiency of 90%. If a standard p-n junction is used as an avalanche diode, to fully deplete such a long region and provide a high field would require a reverse bias of ~500 V. If the doping is constant, then the electric field is linear with depletion distance and only within a small area of the depletion region will the field be high enough for impact avalanche. For the remainder of the depletion width, where the field is lower, the carriers drift through the region at the saturated carrier velocity. The field then must be high enough for impact avalanche and to maintain velocity saturation in the drift region.

Fortunately, the doping profiles of the device can be controlled to allow for a thin avalanche region and a drift region with saturated carrier velocities. Figure 19a shows the schematic of a "reach-through" avalanche

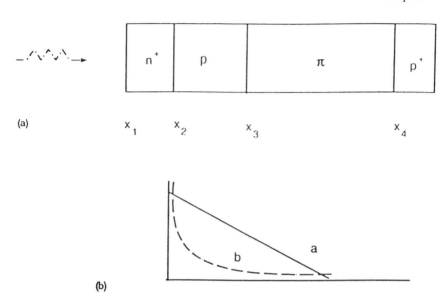

Figure 19 Schematic of an n^+-p-π-p^+ reach-through diode and the electric field across the device for a (a) p-n junction and (b) the reach-through diode.

photodiode. The electric field distribution is shown in Fig. 19b and is compared to that for the one-sided abrupt p-n junction. The doping concentration of the p-region is $\sim 10^{16}$ cm^{-3} and is roughly two orders of magnitude higher than the π-region. The depletion region of the reach-through device extends across the entire device from the n^+- to the p^+-side. Most photons are absorbed in the long π-region. Since, for silicon, electrons are the primary ionizing carriers, these photogenerated electrons are then injected from the p-side for multiplication. The impact process is spread across the n^+-p junction at x_2; then electrons must drift across the p-type region. Avalanche occurs in the high-field region. Figure 20 shows a schematic of a n^+-p-π-p^+ reach-through device.

Control of the avalanche process is governed by the ability to control microplasms, which are small inhomogeneities in the junction which undergo premature breakdown. Microplasms have been virtually eliminated by careful processing and with a specialized design that keeps the active volume as small as possible. Control of the purity of the material and homogeneity is important. Control of the temperature of the device is important in maintaining a uniform threshold voltage. Control of the doping and depth of the avalanche p-region is important, since this is the

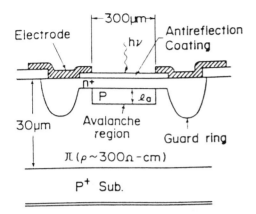

Figure 20   Cross section of a silicon reach-through device. (From Ref. 8.)

region which determines the value of the excess noise factor F. Normally, the p-region is fabricated using ion implantation, which allows for excellent control of the p-region.

Surface breakdown effects can be minimized by a guard ring fabricated surrounding the device to ensure high electric fields never reach the surface. A p^+-channel stop surrounds the detector to prevent surface-layer leakage from conductivity inversion of the surface.

In an idealized reach-through device, another high-resistivity π-region is inserted between the p- and the n^+-regions. This allows for independent control of the width of the p-region and hence better control of the avalanche threshold voltage.

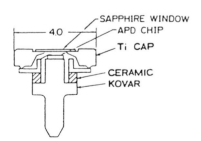

Figure 21   Hermetic package of a avalanche photodiode used for fiber-optic communications. (From Ref. 8.)

The avalanche device has a surface antireflection coating to minimize surface reflections. Because of the high operating voltages and the need for control of APD material properties, the APD must be packaged in a hermetically sealed package that is filled with dry nitrogen. The package is designed to maximize quantum efficiency, and the diode is mounted close to the window to ensure that all light is illuminated on the absorption area (Fig. 21).

8 COMPOUND SEMICONDUCTOR PHOTODIODES

For optical detectors required for operation in the 800–900 nm range and for solid-state applications in the visible region of the spectrum, silicon is the most familiar semiconductor for photodevice applications. The ease of processing and relatively low noise equivalent power for silicon devices often make it the material of choice if it can be use in this spectral region. However, for longer-wavelength applications, the absorption cutoff does not allow its use for wavelengths >1000 nm. Specifically, silicon devices cannot be used for applications in long-haul optical fibers. Because the minimum of attenuation for glass optical fiber exists in a "window" between 1.2 and 1.6 micron, this wavelength range is preferred, and alternative semiconductor materials must be used for fiber-optic detectors. Also, for devices sensitive to the mid- and far-IR, semiconductors may not be practical for those applications.

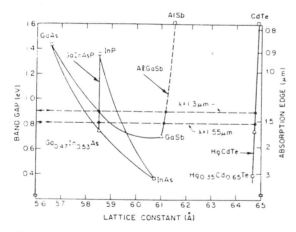

Figure 22 Bandgap energy vs. lattice parameter for compound semiconductor alloys showing the useful range of energies for photodetection. (From Ref. 5.)

The 1.2–1.6 μm range corresponds to band-gap energies centered about 0.9 eV. As seen in Fig. 22, the choice of semiconductor materials other than Ge is limited to $In_xGa_{1-x}As$, $Al_xGa_{1-x}Sb$, and $Hg_xCd_{1-x}Te$. Germanium photodiodes have been limited by the inability to control excess dark currents, and Ge avalanche photodiodes suffer from excess noise problems due to the material having almost equal hole and electron ionization coefficients.

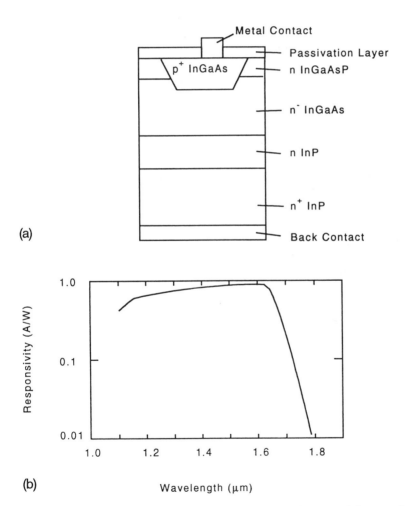

Figure 23 Structure of a GaInAs *p-i-n* diode, and responsivity as a function of wavelength. (After Ref. 15.)

In$_x$Ga$_{1-x}$As *p-i-n* diodes and In$_x$Ga$_{1-x}$As avalanche devices have both been commercially available. Figure 23a shows the structure of a In$_x$Ga$_{1-x}$As *p-i-n* device fabricated on InP substrates. Figure 23b shows the measured spectral dependence of responsivity of such a device.

For fiber-optic applications, a package can be suitably designed where the detector is not background limited. For an In$_x$Ga$_{1-x}$As *p-i-n*, the device has no internal gain and therefore operates where the primary noise source is the thermal noise of the following amplifier. For high-speed (high baud rate) applications, the amplifier following the *p-i-n* diode will be fabricated using compound semiconductor MESFETs (metal-semiconductor field-effect transistors). The integration of a *p-i-n* diode and a hybridized transimpedance amplifier using compound semiconductor MESFET technology is called a PINFET.

Avalanche devices have been suggested in the AlGaSb and HgCdTe systems as well. These latter two semiconductors have favorable band structures to avalanche breakdown. Both have a spin-orbit splitting in the valence band which is on the order of the band gap needed for the 1.3–1.5 μm detecting region. This is favorable as it leads to avalanche diodes with low threshold voltages for impact avalanche multiplication. Al$_x$Ga$_{1x}$Sb must be grown on GaSb substrates, and is lattice matched to these substrates with an addition of a small amount of arsenic [5]. However, material processing problems have greatly limited the usefulness of both Al$_x$Ga$_{1-x}$Sb and Hg$_x$Cd$_{1-x}$Te as semiconductors for near-IR detectors.

For mid-IR both Pb$_x$Sn$_{1-x}$Te [16] and Hg$_x$Cd$_{1-x}$Te photoconductors have been developed. No commercial junction photodiode exists for wavelengths greater than 2 μm. Most applications for detectors in the mid-IR are in defense industries or in infrared astronomy.

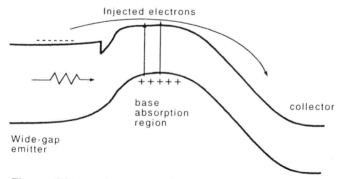

Figure 24 Band structure of a heterostructure phototransistor. The wide-gap emitter acts as a "window" so that absorption primarily takes place in the base.

8.1 Heterojunction Phototransistor

Figure 24 [17] illustrates the operation of a heterojunction phototransistor. A bias is first placed between the collector and the emitter. A wide-bandgap "window" material is inserted to allow the light to penetrate the base of the transistor. At the base, the light is absorbed, creating an electron-hole pair. The holes generated in the base collect there, altering the potential of the base-emitter junction. Decrease of the base-emitter potential causes more electrons to be emitted from the emitter to the base, resulting in an increase in collector current. If the electron transit time is shorter than their recombination time in the base, the result is current gain, as in any normal transistor.

While phototransistors have been fabricated successfully in silicon, their small bandwidths have kept them at a technological disadvantage to silicon APDs. Meanwhile, due to lackluster performance of longer-wavelength avalanche photodiodes, HPT devices have been investigated for applications of >1 μm.

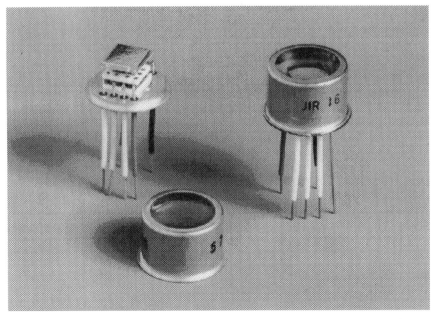

Figure 25 Commercial germanium photodiodes. (Photo courtesy of EG&G Judson.)

REFERENCES

1. G. E. Stillman and C. M. Wolfe, Avalanche photodiodes, in *Semiconductors and Semimetals 12* (R. K. Willardson and A. C. Beer, eds.), Academic, 1977, p. 291.
2. K. L. Anderson and B. J. McMurtry, High speed photodetectors, *Proc. IEEE, 54*, 1335 (1966).
3. R. B. Emmons and G. Lucovsky, An available power-bandwidth product for photodiodes, *Proc. IEEE Corresp., 52*, 865 (1964).
4. H. Melchior, A. R. Hartman, D. P. Schinke, and T. E. Seidel, Planar epitaxial silicon avalanche photodiode, *Bell. Syst. Tech. J, 57*, 1791 (1978).
5. T. P. Pearsall and M. A. Pollack Compound semiconductor photodiodes, in *Semiconductors and Semimetals 22D* (W. T. Tsang, ed.), Academic, 1985, p. 173.
6. S. R. Forrest, Sensitivity of avalanche photodetector receivers for high-bit rate long-wavelength optical communication systems, in *Semiconductors and Semimetals 22D* (W. T. Tsang, ed.), Academic, 1985, p. 329.
7. S. R. Forrest, Optical detectors: three contenders, *IEEE Spectrum*, 76 (May 1986).
8. T. Kaneda, Silicon and germanium avalanche photodiodes, *Semiconductors and Semimetals 22D* (W. T. Tsang, ed.), Academic, 1985, p. 247.
9. P. P. Webb, R. J. McIntyre, and J. Conradi, Properties of avalanche photo-diodes, *RCA Rev., 35*, 234 (1974).
10. S. M. Sze, *Physics of Semiconductor Devices*, 2nd ed., Wiley, 1981.
11. P. A. Wolff, Theory of electron multiplication in silicon and germanium, *Phys. Rev., 95*, 1415 (1954).
12. C. A. Lee, R. A. Logan, R. J. Batdorf, J. J. Kleimack, and W. Wiegmann, Ionization rates of holes and electrons in silicon, *Phys. Rev., 134*, A761 (1964).
13. R. B. Emmons, Avalanche photodiode frequency response, *J. Appl. Phys., 38*, 3705 (1967).
14. R. J. McIntyre, Multiplication noise in uniform avalanche diodes, *IEEE Trans. Electron. Dev., ED-13*, 164 (1966).
15. H. J. Wojtunik, PIN diodes provide low-cost detectors for fiber lasers, *Laser Focus World*, March 1992, 115.
16. H. Melchior, M. B. Fisher, and F. R. Arams, Photodetectors for optical communication systems, *Proc. IEEE, 58*, 1466 (1970).
17. J. C. Campbell, Phototransistors for lightwave communications, *Semiconductors and Semimetals 22D* (W. T. Tsang, ed.), Academic, 1985, p. 389.

5

Semiconductor Image Detectors

1 CHARGE-COUPLED DETECTORS

Charge-coupled devices (CCDs) have virtually revolutionized the field of digital imaging. CCDs have found applications in consumer electronics, astronomy, medical imaging, and computer vision [1,2]. CCDs form an integrated-circuit imager which is low power, lightweight in construction, and inexpensive to produce. An entire TV-compatible solid-state detector can be prepared in pixel sizes of several thousand square on an edge [1] in a chip size less than 100 mm on an edge.

The CCD imaging system is a circuit of monolithic semiconductor devices forming an array of pixels on which the image is focused (Fig. 1). The incident photons release charge in the semiconductor device that can be stored as well as shifted around the array and eventually detected by a charge-sensitive detection system. The shifting is accomplished by a sequence of clock signals applied to the devices. For CCDs used in nonoptical applications, the charge is produced electrically instead of by photons. Nonoptical devices have served as delay lines and storage registers and have found applications in a wide variety of electronics applications, including signal processing, digital filtering, and memory systems.

Optical CCD imaging technology is now capable of responding to wavelengths ranging from the soft x-ray and UV region (~0.1 nm) to the near IR (~900 nm). For imaging applications, picture elements capable of re-

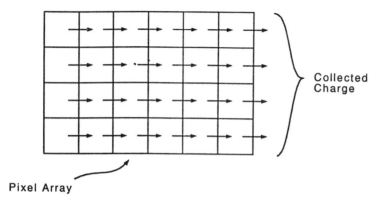

Pixel Array

Figure 1 Photons incident on a CCD array release charge in individual pixels that can be stored or shifted to a charge-sensitive detector.

solving 8 to 12 bits of information, at readout rates in the tens of megapixels, are now possible. New gate and back-illumination technology, and hybrid systems involving other imaging device technology, have allowed extending the sensitivity to short wavelengths, allowing extension into the hard x-ray region.

The heart of the CCD imaging system is the charge-coupled device itself, which is essentially a metal-oxide semiconductor (MOS) capacitor fabricated on p-type silicon. Figure 2 illustrates the energy band diagrams of a MOS capacitor under various bias applications. Although commonly labeled the oxide, the insulating region is either SiO_2 or Si_3N_4 (silicon nitride), and the metal gate may be an appropriate metal such as aluminum or a conducting refractory material such as polysilicon. In an idealized representation of Fig. 2 no surface charge is assumed on the oxide interface between the oxide and the semiconductor, and thus the bands of the semiconductor are flat where they meet the interface with the insulator. In reality, there will be interface states producing surface charge which adds either a positive or negative potential to the interface, the polarity depending on the sign of this surface charge. This charge is due to the presence of impurities or defects at the interface. With no applied bias, this charge is represented by a density of surface states termed the "surface potential," Ψ_S. The sign of the surface potential is dependent on the net sign of the surface charge.

In Fig. 2a no voltage is applied to the gate in the "flat-band condition." In Fig. 2b, a negative voltage is applied and the bands bend according the redistribution of majority carriers (holes). The holes are swept toward the

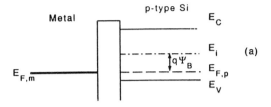

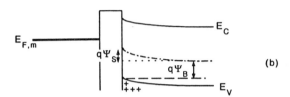

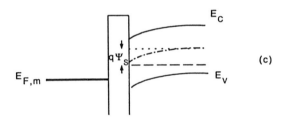

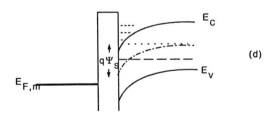

Figure 2 Band bending in an MOS capacitor. (a) So-called flat-band condition with no applied bias and no interface states. (b) With a negative bias, accumulation of holes at the interface results from a variation of the surface potential. (c) Under positive bias, depletion of holes results as holes are swept toward the substrate. (d) Inversion under a large enough positive voltage causes minority carriers (electrons) to congregate at the interface.

interface and away from the bulk semiconductor, and an excess of holes then exists behind the oxide. This is the accumulation mode of an MOS device. In depletion mode (Fig. 2c), a positive voltage is applied to the gate and majority carriers are swept away from the oxide-semiconductor interface, as illustrated. In Fig. 2d, a larger positive voltage results in the Fermi level crossing the intrinsic energy level E_i, and the result is inversion, or the formation of a concentration of minority carriers (electrons as used in this example) at the interface between the oxide and the semiconductor. These minority carriers are separated from the bulk p-type substrate by a depletion region that exists between the bulk holes and the inverted surface electrons. The minority carriers reside in a type of energy "well" formed between the barrier of the oxide and the depletion region of the p-n junction. It is this well in which the photogenerated charge is stored and made available for charge transfer.

Just before the onset of inversion, the device is in depletion mode, and any minority carriers generated by photons will be trapped in the well. These photogenerated carriers can be stored temporarily in the well or transferred out by removal of the bias from the gate.

In depletion mode, the MOS device can be modeled similar to an abrupt p-n junction under reverse bias. The total charge per unit area in the bulk semiconductor, Q_B, is given by integrating the charge per unit volume across the depletion region of width W, or

$$Q_B = -qN_iW \tag{1}$$

where N_i is the concentration of impurities (in the case of the figure, acceptors). Using Poisson's equation, the spatial dependence of the electrical potential $\Psi(x)$ can be determined:

$$\Psi(x) = \Psi_S \left[1 - \frac{x}{W} \right]^2 \tag{2}$$

where Ψ_S is the surface potential defined as

$$\Psi_S = \frac{qN_i}{2\epsilon_s\epsilon_0} W^2 \tag{3}$$

and where $\epsilon_s\epsilon_0$ is the permittivity of the semiconductor. The applied voltage will appear partly over the insulator and partly over the semiconductor:

$$V_{appl} = V_i + q\Psi_S \tag{4}$$

where V_i represents the potential across the insulator:

$$V_i = \frac{Q_B d}{\epsilon_i\epsilon_0} \tag{5}$$

and where d is the thickness of the insulator. As the applied voltage increases, Ψ_S and W both increase.

The capacitance of the device in depletion is a series combination of the capacitance of the insulator $C_i = \epsilon_i \epsilon_0 / d$ and the capacitance of the semiconductor depletion region.

Figure 2 can be used to define accumulation, depletion, and inversion as

$\Psi_S = 0$	flat band
$\Psi_S > 0$	accumulation
$\Psi_B < \Psi_S < 0$	depletion
$\Psi_S < \Psi_B$	inversion

where "weak" inversion is defined as $\Psi_S = \Psi_B$ and strong inversion is defined as $\Psi_S = 2\Psi_B$. When $\Psi_s = 0$, the MOS device is in the ideal (flat-band) condition. Once strong inversion is obtained, the depletion-layer width reaches a maximum, and any further increase in the applied field will be effectively screened by the charge in the inversion layer.

The gradient of the potential $\Psi(x)$ determines the depth of the potential well. This depth can be increased by increasing the oxide capacitance (decreasing the oxide thickness) or by increasing the doping level N_i.

For the nonideal case, the threshold voltage for the onset of inversion can be determined as

$$V_T = \frac{Q_S}{C_i} + q\Psi_B + q\Phi_{MS} + \frac{(Q_i + Q_{SS})d}{\epsilon_i \epsilon_0} \qquad (6)$$

where Φ_{MS} is the difference in the workfunction between metal and semiconductor (if any), Q_i is the localized charge density at the surface of the insulator, and Q_{SS} is the charge density of surface states located at the interface but within the band gap of the semiconductor. Both the charge represented from surface states and that from the insulator will also determine the gate capacitance. This in turn determines the transient response of the device, and the main difference will be apparent in the time constants for the two. The parameter $q\Psi_B$ is dependent on the substrate doping. Any depletion of the doping at the interface (which occurs with some impurities) would result in an additional term.

A surface-channel CCD can be arrayed as a matrix of MOS capacitors and driven with a three-phase clock as shown in Fig. 3. In a three-phase CCD, each pixel is composed of three gates. A positive voltage applied to the gate of any one CCD device will cause the creation of the potential well where the charge created by an incident photon can be stored. On application of the first phase of the clock, every third device is activated and charge is stored in this first set. During the second phase, charge from

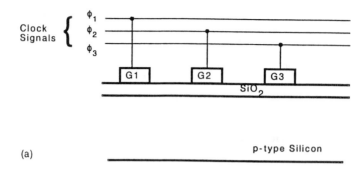

(a)

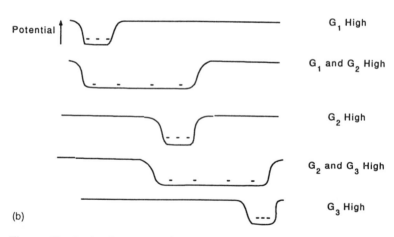

Figure 3 Basic charge transfer process: (a) a CCD driven by an overlapping, three-phase clock. (b) potential wells under each gate during the time each gate is high.

the first set is transferred into the second set of devices, which have been activated by the second clock phase. In the third phase the charge is yet again transferred to the third set of devices and the process repeats itself. Figure 3b illustrates variations of the potential well with applied voltage. The net effect is a linear motion of the charge from the element where it was initially generated to the end of the CCD row where it will be eventually measured by a MOSFET on-chip amplifier. In this manner, an optical image can be recorded by scanning the scene vertically through the imager while rapidly reading out the register in the horizontal rows.

If all the MOS devices in the CCD collected charge, the combined delays of charge moving throughout the array would severely limit the speed at which one image could be processed. Instead, some of the CCDs

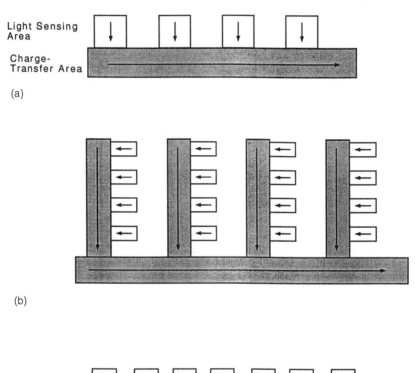

Light Sensing Area

Charge-Transfer Area

(a)

(b)

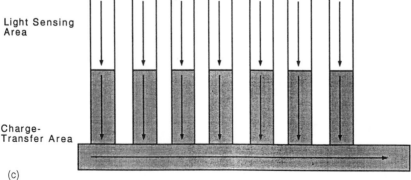

Light Sensing Area

Charge-Transfer Area

(c)

Figure 4 (a) Line, (b) interline transfer, and (c) frame transfer CCD imagers.

do not collect charge but act merely to store or to shift out collected charge. Figure 4 shows a linear imaging array, with optical storage CCDs linked with CCD registers which transfer the stored charge. The gray registers are storage registers, while the white registers are collection registers. Figure 4b shows how linear arrays can be arranged to form a two-dimensional image detector.

A problem with CCDs which have storage elements mixed with image elements is that the presence of the nonimaging elements lowers resolution. Resolution can be improved by using a frame transfer imager. A frame transfer imager consists of an image area, a storage area, and a readout register. The image area uses all of the incoming light and records the image as charge stored in each element. The charge is collected, then moved to the storage section, where the entire frame is temporarily stored before the contents of the frame are shifted out. During the next collection period, the image in the storage frame is transferred one line at a time to a readout register (Fig. 4c). This allows for high frame rates, as the readout registers can be clocked at the conventional TV broadcast rates.

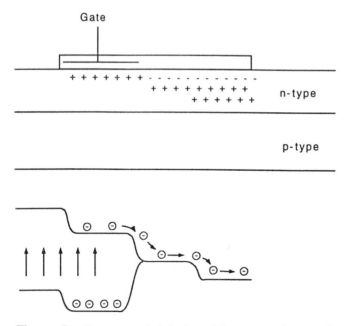

Figure 5   Charge-coupled device with asymmetric potential well for use in a single-phase CCD. Multiple ion implants are used to vary the symmetry of the well. (From Ref. 3.)

A two-phase or single-phase CCD can be made if the potential wells of the device are not symmetric with respect to the direction of the charge transfer. Figure 5 [3] illustrates a single-phase CCD, where different ion implantations are used to produce a step in the potential profiles. Alternatively, two layers of oxide, or two different dielectrics can be used to vary the depth of the well along the line of charge transfer. Figure 5b illustrates the asymmetric potential well of the single-phase device. The variation of the potential assists in the charge transfer process. The single gate varies the depth of the well directly beneath the gate and empties electrons over the potential barriers by reducing the well size under the gate when a voltage is applied to that gate. The well of Fig. 5b created by ion implantation is sometimes called a *virtual* well, and the single-phase or two-phase device is then referred to as a *virtual phase* device.

Unlike three- and four-phase CCDs, two-phase CCDs can operate with non-overlapping clocks. In non-overlapping clocks, called "push" clocks, the signal pushes the charge over the barrier created by the implant.

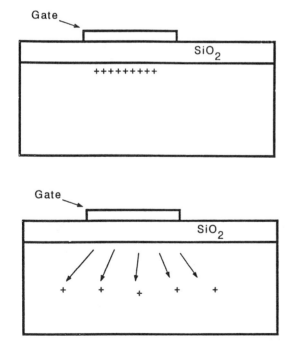

Figure 6 Basic operating function of the charge injection device. Instead of transferring along the surface, charge is measured as a substrate current.

2 CHARGE INJECTION DEVICE (CID)

The basic charge injection device operation is illustrated in Fig. 6. The primary difference between CIDs and CCDs is how the collected charge is measured. In the CCD structure, charge is collected, stored, and shifted out through other CCD registers to be measured by a charge-sensitive amplification system. In the CID device, charge is collected, stored, and discharged through the substrate to be measured as substrate current. The photoinduced charge is collected in the potential wells in the same manner as for CCDs, and for the most part all other functions of the two devices are the same.

CIDs can be used in applications which require a high dynamic range. Dynamic range is increased in a CID as the x-y addressability and the way charge is read through the substrate in CIDs help minimize charge blooming, increasing resolution (see Section 5.3).

3 CCD OPERATING PARAMETERS

Basic parameters used to characterize a CCD include system noise, charge transfer efficiency (CTE), charge collection efficiency (CCE), and quantum efficiency. Most CCD arrays consist of hundreds, if not thousands, of array elements. Thus it is important to minimize the amount of charge loss as charge moves through the array. During the shifting process, charge must transfer over a potential barrier that exists between the array elements. The spacing between individual elements becomes a concern as well as the physical nature of the potential barrier. Many improvements in device design and processing have overcome severe charge loss during transfer, improving efficiencies, and minimizing noise.

3.1 Charge Transfer Efficiency

The charge transfer efficiency, η_{CTE}, is a measurement of the ability of the array to transfer charge without charge loss. It is defined as the fraction of charge transferred from one element to the next. The fraction of charge left behind is the transfer loss, ϵ_{CTE}. If a single charge packet of total integrated charge Q_0 is transferred down a register n times, the net charge remaining is $Q_n = Q_0\eta^n \cong Q_0(1 - n\epsilon_{CTE})$. Original η_{CTE}s were on the order of 99%, while present η_{CTE}s approach or exceed 6–9 s (99.9999%) in efficiency.

Besides charge transfer over the potential well separating elements, another major limitation to the charge transfer efficiency is charge trapping. Surface trapping in surface-channel devices is caused by defects in the silicon/SiO$_2$ interface. These defects are introduced by impurities at the interface, by inhomogeneities, and by the incoherent nature of the interface. Some charge-trapping effects can be minimized by initially filling and maintaining each CCD element with a fixed amount of background charge so that the traps do not interfere with the signal charge. This background charge is often referred to as *fat zero*.

Surface trapping effects can also be minimized by using a buried-channel device. In buried-channel CCDs fabricated on *p*-type substrates, for example, a thin *n*-type region is made directly under the SiO_2. This *p-n* junction forces the potential well away from the Si/SiO_2 interface. Because of the extra *p-n* depletion region, the well depth is shallower for buried channel as for surface channel devices. Buried-channel devices have smother electrostatic fringe fields and switch faster than surface-channel devices, but can hold less charge.

3.2 Mechanisms of Charge Transfer

There are three main mechanisms that enable charge to be transferred from one CCD element to the next: self-induced drift, thermal diffusion, and fringing field drift. Self-induced drift, or carrier diffusion, is a repulsion technique between like charges and is responsible for the majority of charge transfer for conditions of high charge density. The larger the charge to transfer, the higher is the drift field.

Thermal diffusion accounts for another percentage and can dominate at frequencies lower than an upper frequency limit given by [4]

$$f = 5.6 \times 10^7/L^2 \tag{7}$$

where L is the center-to-center electrode spacing measured in microns. The time constant for transfer under thermal diffusion is

$$\tau_{th} = \frac{L^2}{2.5D} \tag{8}$$

where D is the diffusion coefficient of the carrier.

Fringing field drift refers to movement of the charge under the effects of the fringe field surrounding the gate that parallels the direction of charge transfer. This fringe field is at a maximum at the gate edges and is a minimum at the center of the gate.

Transfer loss can be defined as

$$\epsilon_{CTE} = \frac{\Delta N}{N_s - N_{fz}} \tag{9}$$

where ΔN is the net charge loss during a transfer cycle and N_s and N_{fz} are the signal charge and the background charge, or "fat zero," respectively.

Transfer loss can be caused by several parameters, including potential barrier humps between the wells due to finite distance between wells, inadequate time for transfer of charge and trapping at both bulk and interface states. Charges trapped can be released at later (usually inappropriate) times. Transfer loss then is independent of the magnitude of the charge transferred but decreases with increasing clock frequency.

If the loss due to trapping in surface-channel devices is independent of the signal amplitude, the CCD may be operated with an intentional fat zero where the background charge keeps the trapping states filled and thus facilitates charge transfer. Fat zero charge may account for up to 50% of the charge in the well. However, the presence of fat zero also limits the sensitivity in that the presence of the fat zero adds to background noise.

Charge transfer efficiency can be measured by illuminating the CCD with low-energy monoenergetic x-rays [3]. As the absorption length for x-rays is short, most are collected in a very small (within 0.5 μm) region of the surface and can thus be confined to a single point in the CCD array. The x-ray radiation is used to deposit a known amount of charge in a single pixel. As the photon energy is high enough that multiple electron-hole pairs are generated when the x-ray is absorbed, several carriers are generated for each photon and the transfer efficiency for that pixel can be determined.

3.3 Charge Collection Efficiency

Charge collection efficiency is a measurement of the ability of the CCD to collect all the signal generated from a single photon into a single pixel. It is an important parameter in applications which require the confinement of a signal charge without loss to a single pixel element, such as x-ray and UV spectroscopy, and those applications in astronomy requiring high geometric accuracy. For such applications, the charge must be collected in a single pixel without being shared in neighboring pixels.

The degree of charge splitting between pixels depends on where in the pixel the photon is absorbed. Photons that are absorbed close to the front surface of the potential well, where the electric field is high, are considered ideal and are called "single-pixel events." Photons that are absorbed deeper into the device, away from the Si/SiO$_2$ surface, experience a much weaker electric field. These electrons form a "charge cloud" which may diffuse in the substrate of the device and be divided into neighboring pixels. Events of this type are termed "split events." Events producing charge that is eventually lost to trapping or recombination are termed "partial events."

The charge collection efficiency, η_{CCE}, can be determined as [5]

$$\eta_{CCE} = \frac{\eta_E}{\eta_i} \tag{10}$$

where η_i is the ideal quantum efficiency (electrons generated/incident photon) and η_E is a parameter defined by

$$\eta_E = \left\langle \frac{\zeta_{pe}}{P_{se}} \right\rangle \tag{11}$$

where the brackets denote an average. The parameter ζ_{pe} is the number of signal charges generated by one photon and shared by all pixels (an efficiency characterizing the partial event) and P_{se} is the number of pixels which collect the charge generated by a single photon (which represents the split event). An ideal CCD would not generate partial nor split events, and the efficiencies would be such that $\eta_E = \eta_i$.

4 BACK-ILLUMINATED DEVICES

In order to improve sensitivity, CCDs are often thinned and illuminated from the back. Because of their improved sensitivity in the blue and UV region, back-illuminated CCDs are preferred for applications involving shorter wavelengths in ground and space-based applications. Back-illumination proved practical in early aluminum-gate devices, since the aluminum tended to block the incident radiation, but was found necessary for high-sensitive polysilicon gate devices as well. The spectral absorption for polysilicon is such that the polysilicon gates will absorb a significant amount of light for wavelengths shorter than 450 nm. For backside illuminated devices with proper antireflection coatings, quantum efficiencies of greater than 90% can be achieved.

Since the absorption coefficient in Si is very high for short wavelengths, the majority of short-wavelength light will be absorbed close to the surface of the backside for back-illuminated CCDs. Figure 7 [6] illustrates the absorption length (inverse of the absorption coefficient) in silicon for the short-wavelength region of the spectrum. Absorption lengths in the ultraviolet may be on the order of a few hundred angstroms. Short absorption lengths may be a problem for backside illumination because the native oxide which grows on the back surface of the thinned CCD forms a potential well that extends several hundred nanometers into the bulk of the substrate. Fixed charge in this native oxide forms trapping centers which trap electrons generated close to the back surface. These trapped electrons are then lost to recombination. Figure 8 illustrates the band bending that

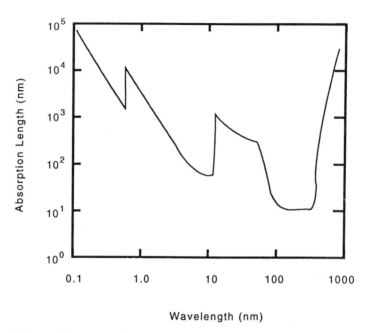

Wavelength (nm)

Figure 7 Absorption length in silicon for short-wavelength light. (From Ref. 6.) For UV and x-ray wavelengths, absorption lengths may only be a few hundred angstroms.

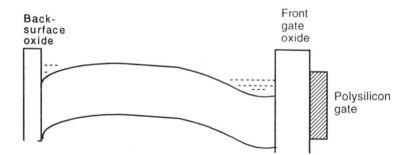

Figure 8 Band bending in a thinned back-illuminated CCD illustrating the additional potential well created from charges in native oxide formed on the back surface.

is believed to exist in thinned CCDs. Back-surface potentials can be reduced by proper backside processing, including implanting a thin p^+ layer close to the back surface in CCDs prepared in p-type substrates. This p^+ layer forms a p^+/p junction which helps minimize the effects of the back surface potential well.

There are other techniques besides implantation used to create the p^+ region on the backside of the CCD. In essence, only a region of negative charge close to the back surface is required to keep that surface in accumulation. The CCD is often prepared by growth of a p^- epitaxial layer on a p^+ substrate. The substrate is then thinned, and different CCD technologies result from how much p^+ material remains. Figure 9 [7] shows how quantum efficiency of short-wavelength light is related to the thickness of the CCD remaining after the thinning process. The thin membrane has a maximum efficiency when converted into a conventional CCD in the "thin" region of the figure. If too much material is removed (overly thin), the efficiency begins to drop. In the "thick" regime several factors result in low efficiency. Included is the aforementioned recombination in the backside potential well between the substrate and the oxide. There also exist interface states between the epitaxial layer and the substrate which cause sites for carrier recombination. The short lifetime for minority carriers in any remaining p^+ substrate material aggravates the back-potential

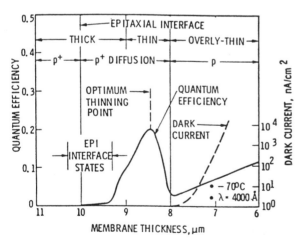

Figure 9 Quantum efficiency of a CCD detecting 400-nm light as a function of the thickness of the CCD remaining after substrate thinning. The figure shows that for commercial devices, an optimum thickness results in the highest QE. (From Ref. 7.)

well as carriers do not have sufficient diffusion times to leave the area close to the back surface.

In the "overly thin" region of Fig. 9, the quantum efficiency begins to drop for the 400-nm light as more p^+ material is removed in the thinning process. Since the depth of the backside well is a function of the doping, this well extends farther into the membrane as more and more p^+ material is removed. Compounding this effect, as excess p^+ material is removed the drift field from the substrate gets smaller. This drift field assists generated carriers in removing from the back to the front surface potential well. For the commercial device, optimum thinning involves thinning the substrate but leaving behind a small portion of the p^+ material which will give the smallest backside potential well and the largest backside drift field.

The technique used to thin usually involves etching the substrate in an immersion etchant. Care must be taken not only to create the proper thickness of membrane but to be sure that thickness nonuniformities are avoided. Thickness nonuniformities cause calibration problems due both to absorption differences in the thickness variations and to backside charging effects due to differing amounts of p^+ material left remaining on the backside.

After thinning, a "flashgate" may be prepared by first passivating the membrane with an oxide at high temperature and then depositing a thin platinum electrode to the back surface. The flashgate has the effect of permanently accumulating the back surface of the CCD due to the work-function difference between the platinum metal and the semiconductor. Alternatively, a transparent conductive coating, such as indium tin oxide (ITO), can be deposited on the back surface and a small negative potential is applied to this electrode to adjust the surface potential.

For CCDs prepared in this manner, quantum efficiencies greater than 60% in the visible and greater than 40% in the near UV can be obtained. However, in the extreme UV and soft x-ray regions, the absorption depth is so short that photogenerated carriers do not have sufficient carrier lifetimes to cross the region to the front surface potential well. Devices sensitive to the EUV and soft x-ray regions can be made by thinning the substrate further, into the "overly thin" regime, to a thickness on the order of the carrier diffusion length, and then using an artificial technique to accumulate the back surface and shorten the back-surface potential well.

An example of such a technique is the backside-charged CCD. The idea is to accumulate an excess of negative charge near the back surface to minimize the back-well depth, without having to leave any p^+ substrate material remaining. This can be done by using the photoemission technique to backside-charge the CCD. Figure 10 shows the energy band dia-

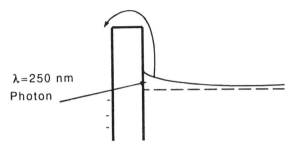

λ=250 nm
Photon

Figure 10 The mechanism behind the backside-charging technique using photoemission. The backside is continuously illuminated with a flood of 250-nm light which causes photoemission of charge from the interface of the back SiO₂ layer over into the back of that oxide where the charge is trapped by absorbed oxygen. (From Ref. 7.)

gram of the photoemission backside-charging technique. The back surface of the structure is continuously flooded with 250-nm-wavelength light. This light promotes photoemission of electrons from the back Si/SiO₂ interface over to the surface of that back oxide. Here the electrons are trapped by trapping states created from oxygen molecules absorbed onto the rear surface. The photoemission causes a depletion of negative charge and hence an accumulation of holes at the rear surface without having p^+ material present. Alternatively, electrons can be removed from the interface by certain gases such as NO absorbed on the rear surface. Electrons are gettered by these gases by tunneling through the oxide [7].

Stern and Catura [8] proposed modeling the back-illuminated CCD by assuming the p^+ region as a "dead" region where photoexcited carriers are assumed lost. In this treatment, the quantum efficiency was simply determined as

$$\eta = e^{-\alpha X_a} \tag{12}$$

where η is the quantum efficiency, X_a is the thickness of the p^+ region, and α is the absorption coefficient. Walker et al. [9] modified this by assuming a uniform electric field E across the shallow p^+ region and hence determined a value for η as

$$\eta = \frac{1}{1 + kT\alpha/qE} - e^{-\alpha X_a} \tag{13}$$

Huang et al. [10] proposed a more extensive model where the efficiency of the CCD is divided into the p^+ region and the p^- region, and the total

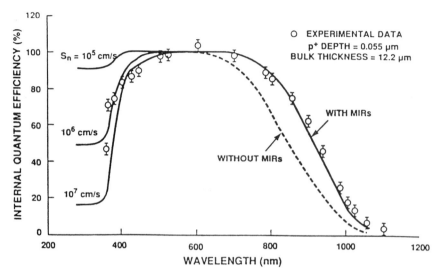

Figure 11 Calculated and measured internal quantum efficiencies for three values of the surface recombination velocity S_n. Effects of multiple internal reflections (MIRS) are also shown. (From Ref. 10.)

efficiency is the sum of these two regions. Huang also proposed a model where multiple internal reflections and the effects of surface recombination could be included.

Results of this model are shown in Fig. 11. Internal quantum efficiencies were determined by measuring external quantum efficiencies and correcting for the reflectance. The theoretical curves were fit to experimental data using different values of the surface recombination velocity S_n. The best fit of the data for S_n is for $S_n = 1 \times 10^6$ cm/s. The figure indicates that allowing for multiple internal reflections is important in determining quantum efficiencies for long wavelengths.

5 NOISE IN CCDs

The smallest size pixel element of a CCD, and hence the highest resolution, is determined by the smallest allowable signal-to-noise ratio for an individual device, and hence depends on the noise of the device. Noise is also an important factor in determining the smallest available charge that can be detected in low-light-level imaging applications.

In surface-channel CCDs the major limitations to noise are due to the surface trapping effects already discussed and to noise associated with a fat zero signal. In buried-channel devices, the noise is mainly limited to variations in dark current, in a fashion similar to that experienced in other photodetection devices.

For CCD imagers there are basically three sources of noise, which are illustrated in Fig. 12. Under high signal levels, the noise is dominated by pixel-to-pixel variations termed *fix-pattern noise*. Noise from such nonuniformities can be controlled by adjusting the voltages of the clocks that control the electric fields of the ununiform pixels. When the fix-pattern noise is properly reduced, the noise is then limited by the shot noise variations in the signal. This noise has the characteristic half-power dependency.

At the lowest signal levels, the device is limited by the "read noise floor," which consists of intrinsic device variations similar to all semiconductor photodetectors, as discussed in Chapter 4. The read noise is determined mainly by background, charge transfer, and amplifier noise.

Background charge arises from dark current, any fat zero present, residual image noise, and luminescence noise.

As noted earlier, a fat zero is often added on purpose to the devices to minimize charge transfer interference from trapping states. There is a trade-off, therefore, of how much fat zero to add without seriously jeopardizing the read noise floor. Residual image noise can occur from overexposure of the CCD, where extra electrons generated during overexposure are trapped at trapping centers at the Si/SiO_2 interface which are normally empty. These electrons are slow to release and will be observed as a residual image during subsequent processing. Residual image noise can

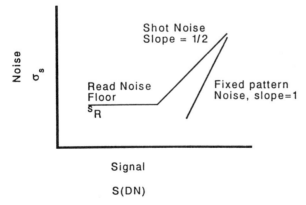

Figure 12 Noise vs. signal level for a CCD imager. The diagram shows three separate regions corresponding to read, shot, and pattern noise.

be controlled in buried-channel devices by inverting the clock signal momentarily so that these trapping states can be emptied.

Luminescence noise is a condition where extraneous long-wavelength photons add charge to the fat zero. These long-wavelength photons are usually generated by an avalanche breakdown condition in a back-biased *p-n* junction on some part of the device.

5.1 Charge Transfer Noise

Charge transfer noise refers to noise which depends on the number of charge transfers that have taken place. Charge transfer noise in virtual phase devices is produced from unwanted charge from the virtual gate region. During charge transfer, holes migrate from the virtual gate and collect under the clocked gate, temporarily inverting it. After transfer, these holes are released at a high energy as the clocked gate returns to its quiescent depleted condition. These high-energy holes tunnel into the oxide, creating electron-hole pairs, and the electrons are then collected under the gate during charge transfer. The more charge transfer that takes place, the larger is the accumulated charge.

5.2 Dark Current

Noise in the CCD is adversely affected by dark currents. In particular, thinning the CCD increases the dark current due to the "hopping conduction" mechanism of these currents. In very-low-doped material, conduction proceeds by a hopping, or tunneling, of carriers between defects in the band gap of the semiconductor or oxide. In this case, the hopping conduction involves the Si/SiO_2 interface states.

Thinning the sample means removing p^+ material; hence the doping in the membrane drops. As it drops, the percentage of thermally generated electrons increases. Decreasing the doping as the sample is thinned also increases the depth of the backside potential well, as explained earlier. Thus, thermally generated carriers have a shorter distance to diffuse to the front surface potential well and be counted along with the signal charge as a dark current.

Dark currents may be suppressed by using a multipinned phase (MPP) [11] technology. In MPP technology, a negative bias is placed on the gate during charge integration, which causes the channel under the oxide to be accumulated with holes. This has the effect of populating interface states with holes and thus suppressing the hopping conduction mechanism. A *p*-type boron implant under one phase of the pixel is used to maintain a potential well during integration.

5.3 Dynamic Range

If the dynamic range of the image is greater than the dynamic range of the imager, then the process of blooming may result. Excess light will create more electrons than what can be stored in the well, and these electrons will spill over into neighboring pixels along the same CCD row or column. The result is a white vertical or horizontal line on the displayed image. The lag of the imager from frame to frame may be a cause of blooming, and not the dynamic range of the CCD itself. If the array is slow to reset, some charge from a previous frame may be left in the current frame. To prevent this type of dynamic range problem, CCDs with high reset rates have been designed.

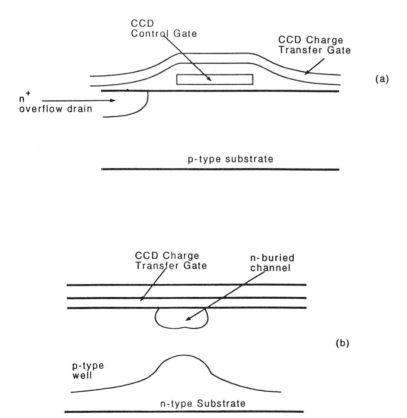

Figure 13 Blooming control devices used in CCDs including (a) an overflow drain and (b) a vertical antiblooming device. (From Ref. 12.)

Blooming control devices can be incorporated into the CCD. Figure 13a shows an example of a control gate and a overflow drain used to control blooming. When the voltage on the control gate is more positive than the voltage on the CCD electrodes, the *p-n* junction formed by the overflow drain will be forward-biased and electrons will spill out the drain. Figure 13b shows a vertical antiblooming device in a frame transfer CCD.

6 PHOTON TRANSFER CURVES [3,5]

The photon transfer technique can be used to evaluate CCD efficiency in the short-wavelength (less than visible) regions. Figure 14 shows the overall transfer function of a CCD, its following amplifier, and A/D conversion. The five blocks are associated with the following transfer functions: (1) photons detected/incident photon (the interacting quantum efficiency QE_i); (2) the electrons generated/incident photon (the internal quantum efficiency η_i); (3) volts/electron transferred (the CCD sensitivity S_V); (4) the signal channel gain, A_1; and (5) the digital number/volt (the transfer function of the A/D converter), A_2. A read noise, $\langle \sigma_R^2 \rangle$ is also present. The eventual digital signal obtained for a given digital number DN, $S(DN)$ is

$$S(\text{DN}) = P \cdot QE_i \cdot \eta_i \cdot S_V \cdot A_1 \cdot A_2 \tag{14}$$

where P is the average number of incident photons per pixel. The efficiency can be related to two parameters, K and J, such that

$$\eta_i = \frac{K}{J} \tag{15}$$

where $K = (S_V \cdot A_1 \cdot A_2)^{-1}$ and $J = (\eta_i \cdot S_V \cdot A_1 \cdot A_2)^{-1}$. The photon transfer technique provides a way to determine the J and K parameters. The major difference between J and K is the parameter η_i, which is found experimentally for x-ray illumination to depend on the incident photon energy E of the radiation by

$$\eta_i = \frac{E \text{ (eV)}}{3.65} \tag{16}$$

Thus for $E > 3.65$ eV ($\lambda < 340$ nm) more than one electron will be generated for each incident photon.

The parameters K and J can be determined by measuring the mean square noise $\langle \sigma_s^2 \rangle$ as a function of signal strength. The noise can be related

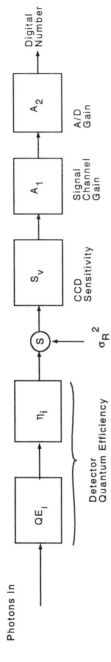

Figure 14 Block diagram of the transfer function of the CCD and its associated electronics.

to the signal by

$$\langle \sigma_s^2 \rangle = \left(\frac{\partial S(DN)}{\partial PI} \right)^2 \langle \sigma_{PI}^2 \rangle + \left(\frac{\partial S(DN)}{\partial K} \right)^2 \langle \sigma_K^2 \rangle + \left(\frac{\partial S(DN)}{\partial \eta_i} \right)^2 \langle \sigma_{\eta_i}^2 \rangle + \langle \sigma_R^2 \rangle$$

(17)

where $\langle \sigma_{PI}^2 \rangle$ and $\langle \sigma_R^2 \rangle$ are the pixel noise and the read noise, respectively. The parameter PI is the number of interacting photons per pixel and equals $P \cdot QE_l$. At minimum signal level, $\langle \sigma_{PI}^2 \rangle = PI$.

For the case of visible light ($\lambda > 340$ nm) only one electron-hole pair is generated for each photon, $\eta_i \approx 1$, and Eq. 14 can be written as

$$S(DN) = PI \cdot K^{-1}$$

(18)

Differentiating Eq. 18 and assuming $\sigma_K^2 = 0$ yield an expression for K:

$$K = \frac{S(DN)}{\sigma_S^2(DN) - \sigma_R^2(DN)}$$

(19)

Likewise, for incident light of $\lambda < 340$ nm,

$$S(DN) = PI \cdot J^{-1}$$

(20)

and the same relation is given for J:

$$J = \frac{S(DN)}{\sigma_S^2(DN) - \sigma_R^2(DN)}$$

(21)

The parameters J and K can thus be found by measuring the signal and its variance for two different light sources. The photon transfer curve is the plot of noise $\langle \sigma_S \rangle^2$ as a function of signal $S(DN)$. For example, Fig. 12 shows a photon transfer curve for $\lambda = 700$ nm. $S(DN)$ is the average signal level of the pixels with the array uniformly illuminated. The variance $\sigma_S(DN)$ can be found by subtracting pixel-by-pixel two frames illuminated at the same light level, calculating the standard deviation of the result. The level of illumination is increased until the shot noise component of σ_s dominates over the read noise σ_R, and the slope of the shot noise component is determined as $1/2$. From the graph the read noise σ_R can be determined, and the intersection of the slope of the line of the shot noise component with $\sigma_S = 1$ yields the parameter K.

Figure 15 [5] illustrates photon transfer curves for illumination from different light sources on the same CCD. For the 700-nm source, a value of K of 2.3 electrons/DN was determined. For the wavelengths of 2.1 Å and 1216 Å, values for J can be determined and used with the measured value of K to find η_E values of 1470 electrons/pixel and 3 electrons/pixel, respectively.

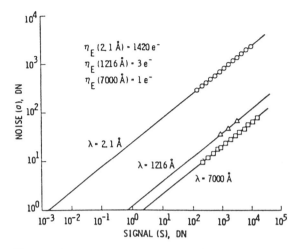

Figure 15 Experimental photon transfer curves taken at three different photon energies. Each yields a different effective quantum yield. (From Ref. 5.)

7 DETECTOR ARRAYS FOR IR/EUV

7.1 Multichannel Plate Detectors

The sensitivity of a CCD detector can be increased by using a front-end amplification device called a multichannel plate (MCP). The MCP is sensitive to input stimuli from mid and extreme UV, soft x-rays, and charged particles such as ions and electrons having energies from 10 to 100,00 eV, thus increasing the range of the CCD into the subangstrom wavelengths.

MCPs can detect ions, electrons, and photons, convert them into electrons, and amplify these electrons by means of a cascade of secondary electron emission. MCPs are plates with an array of small parallel channels or thin tubes of 6–25-μm diameter etched into the plate. Each channel acts as an electron multiplier. The interior of the hole is coated with a film such as CsO which lowers the workfunction of the material. Figure 16a illustrates the band bending at the interior surface of each channel due to the film's presence. Figure 16b shows how electrons cascade down the interior of the channels of the MCP. As they cascade, they impinge upon the surface of the channel where the workfunction is low, resulting in the emission of more secondary electrons. At the end of the channel is an array detector for detecting the electron cascade. A simple array detector may be formed from a multianode array of conductive metal strips. Alternatively, the electron cloud emanating from the channel may

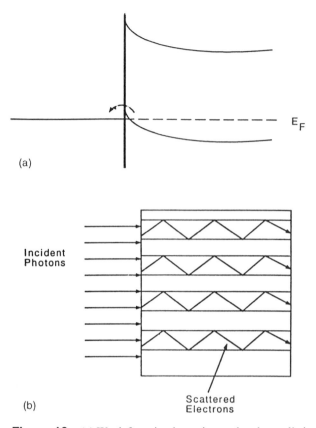

(a)

Incident
Photons

(b)

Scattered
Electrons

Figure 16   (a) Work function lowering at the channel's interior surface. (b) How electrons cascade down a channel within a multichannel plate.

be detected by a phosphor screen, thus converting the signal back to an optical image which can be read by a CCD.

The MCP is often coupled to the CCD array through a fiber-optic channel, as illustrated in Fig. 17. The multichannel plate converts the incoming light to a stream of electrons, which are converted back to photons by the phosphor screen. The photons are then channeled to the CCD by the fiber-optic couplers. If the MCP generates M electrons for each incoming photon, and, in turn, a fraction X of these electrons will generate photons from these electrons, a gain of MX is realized for this device. Values of the gain can range from 10^3 to 10^8, and thus single-photon-counting techniques can be implemented.

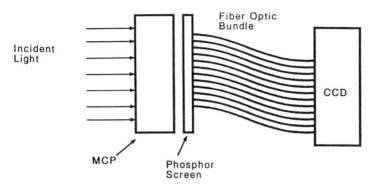

Figure 17 High-sensitivity image detector using fiber-optic-coupled multichannel plate (MCP) technology. The MCP converts a photon image to a stream of electrons, which is converted to a group of photons (thus realizing internal gain) to be detected by the CCD.

Another type of front-illuminated CCD sensitive to short wavelengths is a scintillation device where the front surface is coated with a phosphor sensitive to the blue or UV regions. After the initial absorption of short-wave light, this phosphor reemits photons in the visible region where the CCD is more sensitive [3]. A problem with the scintillation device is that overall efficiency may be limited by the efficiency of the phosphor.

7.2 MWIR Detector Arrays

Because of the limited spectral range of silicon, detector arrays for mid- and far IR must rely on a photodetection process other than that utilized for the CCD. If a photovoltaic device is needed for a particular application, a low-band-gap material is necessary. Focal plane array (FPA) detectors of different materials with small band gaps have been fabricated for use in the 1–30-μm spectral range. Such array detectors have use in military and infrared astronomy applications.

Unlike images from visible sources, images from infrared sources have low contrast and a large background component. An infrared imager must have a high dynamic range in order to be sensitive to the low contrast and not be saturated by the large background. If the background can be removed during the integration process, the demands on dynamic range can be lowered. To successfully remove the background requires that the pixel uniformity of the detector be maximized [14].

Photovoltaic and photoconductive detectors for infrared suffer from several intrinsic noise problems. Because of the low band gap energy, the intrinsic conductivity of an IR detector will have an appreciable amount of thermal generation and thus a high dark current. Thermal noise from incident background thermal sources must also be dealt with. Because of these two problems, photovoltaic and photoconductive IR devices must generally be cooled to low (liquid nitrogen) temperatures for operation.

Another factor of the low band gap and the high intrinsic conductivity of intrinsic-IR materials is the difficulty in isolating devices from one another on a wafer. The high conductivity of the IR substrate creates a natural channel for diffusion and drift currents so that a high amount of "crosstalk" exists in an array formed this way. It is not possible to create "semi-insulating" material to isolate such devices. Also, large wafers of infrared-sensitive materials are not readily available due to crystal growth difficulties. If devices formed monolithically on a wafer of IR material cannot be isolated from one another, and a large enough array cannot be fabricated on what wafers are available, the only alternative is to use discrete detector devices arranged in an array that must be fabricated by physically inserting or bonding the devices to a substrate. This method of fabrication of the FPA tends to limit its resolution.

Also of concern are materials problems with certain low-band-gap materials. The semiconductor alloy HgCdTe is unstable at even moderate processing temperatures and is usually only usable in a preferred conductivity (p-type). Thus, forming p-n junctions of the device is difficult. Forming MIS structures is not feasible as oxides of IR materials and deposited films contain an unacceptable concentration of interface states which tend to prevent modulation of the band energies, and thus prevent proper device operation.

In spite of these problems, FPAs have been fabricated which are sensitive to the IR regions. These regions are usually separated into the short-wavelength IR (SWIR) 1–3 μm, the medium-wave IR (MWIR) of 3–5 μm, and the long-wave IR of 8–20 μm. Infrared FPAs have been made from PbSnTe, InSb, InAsSb, HgCdTe, silicon doped with various elements to make extrinsic photoconductors, and PtSi detectors. These FPA structures are usually fabricated by mounting the detectors on a silicon CCD that is used as a multiplexor and charge storage device. Front surface-mounted devices have a wire or conductive strip that must pass from the front of the detector to the multiplexer through a via or over the edge. A backside-illuminated detector can be mounted on the surface of the CCD multiplexer by using a metal-bonding process. Figure 18 [15] shows a HgCdTe focal plane array where backside-illuminated devices have been mounted on a CCD with indium "bumps." The detectors are made using

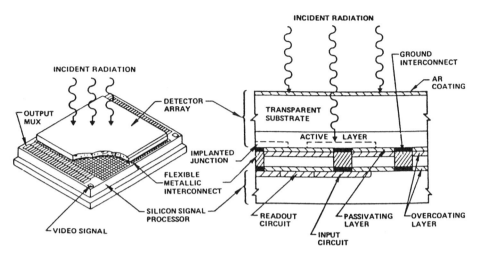

Figure 18 Architecture of a planar hybrid focal plane array (From Ref. 15.)

epitaxial growth on a transparent substrate or by mechanically thinning a bulk device until it is within the diffusion length of the photoexcited carrier. Ion implantation is then used to dope the *p-n* photodiodes.

7.3 Schottky Barrier IR Imagers

Infrared imagers can also be fabricated from Schottky barrier arrays. Schottky barrier arrays have the major advantage in that alternative semiconductors are not used; thus the advantages of silicon processing are maintained.

The imaging device consists of an array of Schottky diodes fabricated on a silicon substrate. The Schottky electrodes are either metals or metal silicides such as PtSi or Pd_2Si which are deposited as a metal then compounded in a solid-state chemical reaction. Figure 19 shows a band diagram of the Schottky imaging process. A metal is deposited on the silicon surface and the device is illuminated through the semiconductor. The long-wavelength photons are absorbed in the metal, and the electron is excited over the metal-semiconductor energy barrier.

The Schottky array operates in frame imaging mode, in a system similar to but slightly different from the frame transfer mode of a CCD. The array is initially disconnected from the bias source, and the Schottky detectors act as floating capacitors to detect the signal. The array senses the optical signal by the storage of the electron in each capacitor, the charge of the

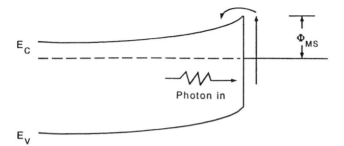

Figure 19   Energy band diagram of a Schottky barrier photodetector. The photocarrier is excited from the metal to the semiconductor, where it is recorded as a fixed charge.

stored electron being opposite that of the depletion-layer charge of the metal-semiconductor junction. At the end of the integration each pixel is recharged back to its original condition, and the amount of current necessary to complete this recharge is measured. This recharge current represents the video signal. For a vidicon-style array, multiplexing and readout is accomplished by a scanning electron beam.

As the radiation must pass through the semiconductor, the semiconductor acts as a low-pass filter, filtering radiation shorter than ~ 1 μm. The long-wavelength cutoff is determined by the metal-silicon barrier height, Φ_{MS}, which is equal to the difference of the work functions of the two materials. Thus, the spectral sensitivity of the detector can be determined by choosing a metal with an appropriate workfunction.

The device must be cooled in order to lower background currents and to prevent thermal emission over the Schottky barrier. PdSi Schottky barrier detectors have relatively low quantum efficiencies, but, because of their low read noise floor, infrared CCD combination using PdSi gates can operate at close to room temperature [16,17].

REFERENCES

1. M. M. Blouke, D. L. Heidtmann, B. Corrie, M. L. Lust, and J. R. Janesick, Large area CCD image sensors for scientific applications, *Proc. SPIE, Solid State Imaging Arrays, 570,* 82 (1985).
2. B. Munier, P. Prieur-Drevon, and J. Chabbal, Digital radiology with solid-state linear x-ray detectors, *Proc. SPIE, Charge Coupled Devices and Solid State Sensors II, 1447,* 44 (1991).

3. J. R. Janesick, T. Elliott, S. A. Collins, M. M. Blouke, and J. Freeman, Scientific charge coupled devices, *Opt. Eng.*, *26*, 692 (1987).

4. W. F. Kosonocky, Charge coupled devices—an overview, in *1974 Western Electron. Show and Conven. Tech. Papers 18*, 2/1 (1974).

5. J. Janesick, K. Klaasen, and T. Elliott, CCD charge collection efficiency and the photon transfer technique, *Proc. SPIE, Solid State Imaging Arrays*, *570*, 7 (1985).

6. M. M. Blouke, W. A. Dalamere, and G. Womack, A simplified model of the back surface of a charge-coupled device, *Proc. SPIE, Charge Coupled Devices and Solid State Sensors II*, *1447*, 142 (1991).

7. J. Janesick, T. Elliott, T. Daud, J. McCarthy, and M. Blouke, Backside charging of the CCD, *Proc. SPIE, Solid State Imaging Arrays*, *570*, 46 (1985).

8. R. Stern and R. Catura, *Proc. SPIE, Charge Coupled Devices and Solid State Sensors II*, *627*, 583 (1986).

9. J. W. Walker, B. H. Breagead, L. J. Hornback, C. G. Robers, K. P. Stabbs, and D. R. Collins, in *Proceedings International Conference on the Application of Charge Coupled Devices*, Naval Ocean Systems Center, San Diego, 1978, p. 141.

10. C. M. Huang, B. B. Kosicki, J. R. Theriault, J. A. Gregory, B. E. Burke, B. W. Johnson, and E. T. Hurley, Quantum efficiency model for p^+-doped back-illuminated CCP imager *Proc. SPIE, Charge Coupled Devices and Solid State Sensors II*, *1447*, 156 (1991).

11. J. Janesick, T. Elliott, R. Bredthauer, C. Chandler, and B. Burke, *Proc. SPIE, X-ray Instrumentation in Astronomy*, *982*, 70 (1988).

12. M. G. Collett, J. G. C Bakker, L. J. M. Esser, H. L. Peek, M. J. H. van de Steeg, A. J. P. Theuwissen, and C. H. L. Weijtens, High density frame transfer image sensors with vertical anti-blooming *Proc. SPIE, Solid State Imaging Arrays*, *570*, 27 (1985).

13. R. A. Bredthauer, C. E. Chandler, J. R. Janesick, T. W. McCurin, and G. R. Sims, Recent CCD technology developments, in *Instrumentation for Ground-based Optical Astronomy: Present and Future* (L. B. Robinson, ed.), Springer, Berlin, 1988, p. 486.

14. J. A. Hall, Problem of infrared television camera tubes vs. infrared scanners, *Appl. Opt.*, *10*, 838 (1971).

15. J. P Rode, HgCdTe hybrid focal plane, *Infrared Phys.*, *24*, 443 (1984).

16. F. D. Shepherd Jr., A. C. Yang, S. A. Rooslid, J. H. Bloom, B. R. Capone, C. E. Ludington, and R. W. Taylor, Silicon Schottky barrier monolithic IRTV focal planes, in *Advances in Electronics and Electron Physics 40b*, Academic, New York, 1976, p. 981.

17. W. F. Kosonocky, Progress in Schottky-barrier IR imagers, *Proc. SPIE, Infrared Systems and Components*, *750*, 136 (1987).

6

Nonlinear Optical Switching Devices

1 INTRODUCTION

An optical switching device may be defined as one that either diverts an optical beam from its path or that may be used as a source for optical modulation. An optical modulator would be used to modulate an optical signal, for example, to encode a digital signal onto a light beam.

An optical switching device may be fabricated as a discrete component or incorporated into an optical waveguide. It must, ideally, consume low power and be operable at a comparatively low optical power level and high bandwidth. Once a switching device is achieved, it is a short step to optical digital logic gates, as well as other optical subsystems, which combined with optical memory components and optical waveguides could lead to an all-optical processor. In the meantime, optical switching devices can find other applications in subpicosecond switching, as it is much easier to generate and propagate subpicosecond optical pulses than subpicosecond electrical pulses.

Optical switching devices are often designed with a long-term goal: the replacement of the electrical transistor with a large-bandwidth all-optical device. Such a device would have gain and have all optical inputs and outputs. Purely photonic devices can be envisioned as having an input and output transfer function with little or no electronic component; such an electronic component may increase the risk of failure or lower bandwidth.

A nonlinear optical device makes use of the nonlinear optical properties of certain materials. The electrical polarization of a material excited by the electric field from a beam of light can be expanded as a power series

$$P_i = \chi_{ij}^{(1)} E_j + \chi_{ijk}^{(2)} E_{jk} + \chi_{ijkl}^{(3)} E_{jkl} + \cdots$$

where the subscripts refer to the Cartesian coordinate system. The higher orders of χ are called the nonlinear susceptibilities and can be dependent upon the crystal symmetry. Normally, only the lowest (linear) order $\chi^{(1)}$ dominates and is the only important term at low intensities. The $\chi^{(2)}$ term is linear with field and multiplies the incident optical field, creating second-harmonic frequencies. The $\chi^{(3)}$ term is quadratic with field and produces third-harmonic frequencies or induces a change in the index of refraction or absorption. The optically induced refractive index change, or absorption change, with light intensity from the third-order nonlinearity can be exploited to make ultrafast optical switches.

2 OPTICAL BISTABLE DEVICES

An optical bistable device has optical inputs and outputs and may or may not include an intermediary electrical component, although a common goal is to produce an all-optical device. An optical bistable device has two possible output states for the same optical input state. The choice between output states is dependent on the history of the input state. It is best described using the setup of Fig. 1. Two parallel but partially reflecting mirrors form a Fabry-Perot etalon. The medium inside the etalon is a nonlinear optical material. Such an etalon could be made, for example, from a nonlinear semiconductor with two cleaved polished faces. The light partially reflects from each face, back into the semiconductor, which is the nonlinear medium. We can assume initially that the frequency of the light and the properties of the etalon are such that the etalon is either tuned to resonance or detuned. Light entering from the left is split and partially reflected by the two mirrors. If the cavity is not in resonance, then the transmitted light intensity I_T is simply proportional to the incident intensity I_I.

The nonlinear nature of the medium, however, dictates that I_T is a nonlinear function of I_I; i.e., $I_T = I_T(I_I)$ is nonlinear. There are two basic types of optical nonlinearities: absorptive and dispersive. In an absorptive nonlinearity, the absorption coefficient of the medium is a function of the incident light intensity, and the medium may become more transparent or more opaque with increased light intensity. If the medium is such that

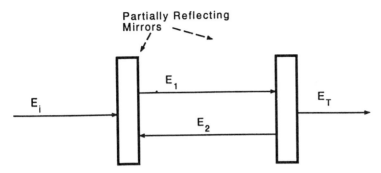

Figure 1 Fabry-Perot etalon as used as an optical bistable device.

it becomes more transparent with incident light, that is, the absorption edge shifts to shorter wavelengths with increased incident light intensity, the medium is often called a *saturable absorber*. A medium where the absorption edge shifts to longer wavelengths with increasing light intensity is a *saturable transmitter*.

In a dispersive nonlinear medium, the refractive index of the medium is dependent on the incident light intensity. Hence, the optical path length of the cavity is also dependent on I_t. Absorptive and dispersive effects usually occur simultaneously and cannot be separated, but at certain wavelengths in practical nonlinear optical media one or the other will tend to dominate.

For the absorptive medium, a state equation can be written for the etalon as

$$E_i = E_t + \Gamma v \tag{1}$$

where E_i and E_t are the incident and transmitted electric field vectors, respectively, and the parameter Γv represents the portion of the field lost (absorbed) in the medium. If v is normalized in such a way as $v/E_t \to 1$ as $E_t \to 0$, then $\Gamma = \alpha L/T$, where α is the (nonlinear) absorption coefficient for the material, L is the length of the cavity, and T is the transmission coefficient of the mirror. For a purely dispersive medium, a similar equation can be written [1]:

$$E_i = E_t + i(\Gamma u + \beta E_t) \tag{2}$$

where u is similar to v, only written in terms of a nonlinear refractive index, and β is a tuning parameter.

The value of Γ is a fundamental parameter of a bistable device, since it represents the finesse of the cavity and determines whether the cavity is tuned to resonance. Figure 2 shows the transfer characteristics of an "ideal theoretical" all-optical bistable etalon as a function of the parameter Γ. For low values of Γ the relationship between I_T and I_I is sublinear. The material absorbs a portion of the incident light intensity, and subsequently the transmitted intensity is reduced. What separates absorption in the etalon from a more familiar absorption process, however, is the presence of feedback. As the value of $\alpha L/T$ becomes increasingly smaller, the curve in Fig. 2 becomes S-shaped, and the negative slope of the curve represents an unstable condition. Hence, for small Γ there is a range of values of the incident intensity for which the output is bistable or has two possible outputs. A monostable device can be made for relatively large Γ as shown in Fig. 3A, and a bistable device for small Γ shown in Fig. 3B. For the monostable device, a large increase in the output intensity is realized as the etalon is tuned into resonance. The tuning can occur by triggering the device with a separate optical source, by tuning to a slightly different wavelength, or by changing the optical intensity of the incident beam. The effect is the optical equivalent of the transistor.

For the optical bistable device, the output depends on the history of the input. As the input intensity is scanned from a low to high value, the output follows a path different from the path it follows as the incident intensity is subsequently reduced. The resulting hysteresis is a result of the unstable range of outputs experienced for low values of Γ.

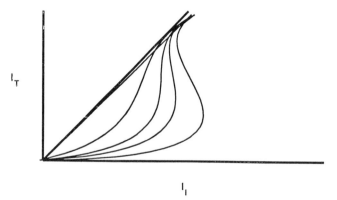

Figure 2 Optical transfer curves of output intensity vs. input intensity for differing values of the parameter $\alpha L/T$. The curves become more S-shaped with decreasing values of $\alpha L/T$.

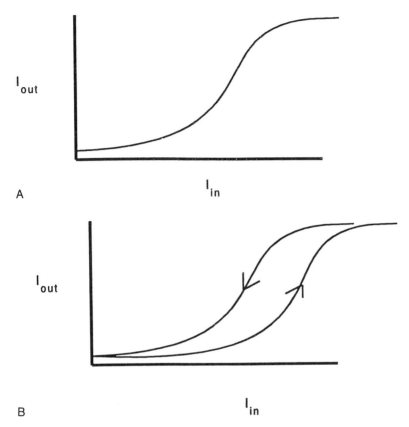

Figure 3 Optical transfer curves of output intensity vs. input intensity for (A) a monostable and (B) a bistable optical etalon.

The phenomenon of optical bistability within an etalon is a combination of the nonlinear energy transfer of the medium with the positive feedback of the mirrors. There are some instances, described later, where the presence of the etalon is not a requirement to achieve bistability.

2.1 Simple Phenomenalistic View of Optical Bistability in an Etalon

We will use as an example the case of an etalon containing a saturable absorber. In the initial case, which is illustrated in Fig. 4, the medium is absorbing or "unbleached." The medium absorbs at the wavelength of incident light, and the cavity is untuned to resonance. The "finesse" of

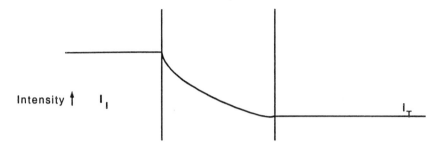

Intensity $\uparrow$ I_I

Figure 4   Fabry-Perot etalon containing a saturable absorber where the medium is unbleached. The intensity of the light is much less than the intensity necessary to bleach the medium. Transmission of the cavity is low, absorption is high, the etalon is untuned (finesse is spoiled).

the cavity is spoiled due to the absorption by the medium. The relative cavity intensity $I_C(z)$ and the transmitted intensity $I_T(z)$ are simply

$$I_C(z) = I_I T e^{-\alpha z} \tag{3}$$

and

$$I_T(z) = T^2 I_I e^{-\alpha z} \tag{4}$$

where T is the transmission and α is the absorption coefficient. Now as the incident intensity I_I is slowly increased, a point is reached where the cavity intensity equals the saturation intensity of the nonlinear medium I_S, and the medium bleaches. At this point the cavity hits resonance and the transmitted intensity suddenly increases (as shown in Fig. 5) to

$$I_T = I_I T^2 \tag{5}$$

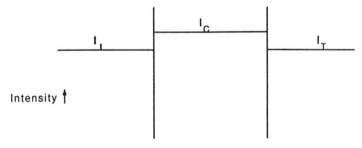

Intensity $\uparrow$

Figure 5   Fabry-Perot etalon containing a saturable absorber where the medium is bleached, the cavity is tuned to resonance, and the transmission of the cavity is high.

and the intensity in the (now bleached) cavity becomes

$$I_C = \frac{I_T}{T} \tag{6}$$

At this point, the intensity in the cavity increases due to constructive interference of the coherent light within the cavity. The transmitted intensity then increases close to 100% of the incident intensity.

If the incident intensity is now reduced, the intensity of the light in the medium will remain above the saturation threshold, that is $I_C > I_S$, and the etalon remains in resonance until the incident intensity is reduced to where $I_C < I_S$.

For the case of I_I increasing, the overall intensity of I_C was lower than it was for I_I decreasing due to the absorption that was then present in the cavity. The point at which $I_C = I_S$ will be different for the two cases, and the output does not follow the same path for increasing/decreasing illumination intensities.

For a dispersive medium, the refractive index depends on intensity, and the Fabry-Perot cavity of Fig. 1 will tune or detune in and out of resonance depending on the intensity of the incident light. Figure 6 shows the transmission characteristics of the cavity as the wavelength of the incident light is scanned. Variation of the refractive index will alter the cavity's optical path length and shift these peaks to the left or right. The peaks may thus be shifted toward or away from an operating laser wavelength. Thus, the etalon can be switched from a highly opaque to a highly transmitting state, or vice versa, by variations in the incident light intensity.

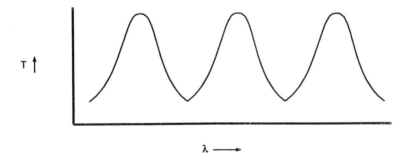

Figure 6 Transmission peaks from a Fabry-Perot etalon with a dispersive medium as a function of the incident light wavelength. The peaks shift as a function of light intensity, and the etalon becomes highly transmitting as a peak nears the laser wavelength where the cavity is in resonance.

Dispersive bistable devices are desirable for the high switching speeds of the effect and low-noise characteristics as compared to purely absorptive devices. Furthermore, the reflective signal from the etalon can be used to generate negative logic.

2.2   Model for Optical Bistability

As stated earlier, the onset of bistability is dependent on the nonlinear nature of the optical medium coupled with positive feedback. A rigorous demonstration of bistability requires a rather intense study of the nonlinear properties of semiconductors derived from quantum-mechanical calculations of absorption and refraction using many-body techniques. Instead, a rather simplistic demonstration of the curves leading to Fig. 2 can be obtained from the analysis of a two-level electron system of a nonlinear medium inside a Fabry-Perot cavity. Extending the relatively simple analysis of this semiclassical model reveals equations relating polarization, dispersion field, and the electromagnetic field.

We start by solving the Schrödinger equation for a one-atom, two-level system. The atoms in the medium are described by an index μ. There are N electrons in our one-dimensional system, and each is elastically coupled to each atom and the displacement from equilibrium is given the value x_μ. As the photons excite the atoms, the electrons become displaced and each contains a contribution to the polarization, which can be described by

$$P_\mu = -qNx_\mu \tag{7}$$

The macroscopic polarization P then can be determined from combining the individual atomic polarizations P_μ. From classical physics, x_μ can be determined by considering the equation of motion of an anharmonic oscillator:

$$\ddot{x}_\mu + \omega_0^2 x_\mu + \sigma\dot{x} = \left(-\frac{q}{m}\right) E(x, t) \tag{8}$$

where $\omega_0^2 = f/m$ and f is Hook's constant. Losses are accounted for by a frictional force $\sigma\dot{x}m$. This model is appropriate as the anharmonic nature of the electron oscillators about the atomic cores as driven by the high field strength of the laser must be accounted for. Using a sinusoidal electric field given by

$$E(z, t) = E_0(z)e^{i\omega t} + \text{c.c.} \tag{9}$$

where c.c. represents the complex conjugate of the first term, a solution for x_μ can be found:

$$x_\mu = \frac{-(q/m)}{\omega^2 - \omega_0^2 + i\omega\sigma} E(x_\mu, t) \tag{10}$$

where x_μ is the equilibrium atomic position. The individual atomic contribution to the polarizability then is

$$P_\mu = \chi E(x_\mu, t)N \tag{11}$$

where the susceptibility χ is given by

$$\chi = \frac{-(q^2/m)}{\omega^2 - \omega_0^2 + i\omega\sigma} \tag{12}$$

In this model, we have assumed the susceptibility to be a scalar. The parameter χ is a complex quantity; the imaginary part represents the absorption as a function of frequency.

In addition, the problem of the oscillating charge can be solved with quantum-mechanical methods. In this fashion, the classical atomic polarization P_μ is replaced by the expectation value p:

$$p = \int \Psi^* (-qx)\Psi \, d^3x \tag{13}$$

The Schrödinger equation is solved with the Hamiltonian

$$H = H_0 + qxE(t) \tag{14}$$

where H_0 is the Hamiltonian of the unperturbed operator and the second term represents the perturbation. The unperturbed wavefunction is determined from

$$H_0\phi_1 = W_1\phi_1 \tag{15}$$

and

$$H_0\phi_2 = W_2\phi_2 \tag{16}$$

The atomic polarization can be determined as

$$p = c_1^* c_2 \exp[-i\omega_0 t]v_{12} + c_2^* c_1 \exp[i\omega_0 t]v_{21} \tag{17}$$

where ω_0 represents the transition frequency between the two levels and is given by

$$\omega_0 = \frac{E_2 - E_1}{\hbar} \tag{18}$$

and v_{jk} is a matrix element represented by

$$v_{jk} = \int \phi_j^*(q\xi)\phi_k \, d^3\xi \tag{19}$$

There is also a second differential equation that can be formed from the population inversion represented by

$$d = |c_2|^2 - |c_1|^2 \tag{20}$$

The value of d represents the difference in the population probabilities (the inversion) of levels 1 and 2. Summing the individual polarizabilities and inversions using

$$P(x, t) = \sum_{\mu} \delta(x - x_\mu) p_\mu \tag{21}$$

and

$$D(x, t) = \sum_{\mu} \delta(x - x_\mu) d_\mu \tag{22}$$

results in differential equations for the macroscopic polarization and the macroscopic inversion:

$$\frac{\partial P}{\partial t} = \frac{|\nu|}{\hbar} ED - [\gamma_\perp + i(\omega_0 - \omega)]P \tag{23}$$

$$\frac{\partial D}{\partial t} = \frac{|\nu|}{2\hbar} (EP^* + E^*P) - \gamma_\| \left(D - \frac{N}{2} \right) \tag{24}$$

These are the so-called Maxwell-Bloch equations describing the polarization of material. The parameters $\gamma_\perp$ and $\gamma_\|$ are the inverses of the atomic relaxation times, N is the number of levels, E is the slowly varying envelope function that represents the electric field, P is the macroscopic atomic polarization, and D is one-half the difference between the populations of the lower and upper levels, as determined by Eqs. 20 and 22.

In order to avoid interaction of standing waves, a ring cavity is assumed as shown in Fig. 7. In this manner we only deal with waves traveling in

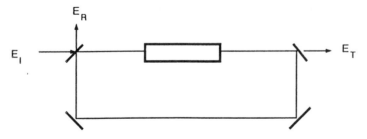

Figure 7   Ring cavity experimental setup used in the model.

one direction. Starting with the wave equation

$$\nabla^2 E - \mu\sigma\dot{E} + \mu\epsilon\ddot{E} = \mu\ddot{P} \tag{25}$$

and applying the slowly varying amplitude approximation to obtain

$$\frac{\partial E}{\partial t} + c\frac{\partial E}{\partial x} = \frac{i\omega_0}{2\epsilon_0}P \tag{26}$$

We obtain the boundary conditions for the wave at mirror 2,

$$E_T(t) = \sqrt{T}\,E(L, t) \tag{27}$$

and at mirror 1,

$$E(0, t) = \sqrt{T}\,E_I + R\,\exp[-i\delta_0]E(L, t - \Delta t) \tag{28}$$

where T and R are the transmissivity and reflectivity, respectively, and δ_0 is the cavity detuning:

$$\delta_0 = (\omega_c - \omega)\frac{2(L + I)}{c} \tag{29}$$

Here ω_c is the frequency of the cavity nearest to resonance with the incident light, and Δt is the time the light takes to travel from mirror 2 to mirror 1 $= (2I + L)/c$.

The steady state is defined when $\partial E/\partial t = \partial P/\partial t = \partial D/\partial t = 0$. From the field equation we find that

$$\frac{\partial E}{\partial x} = -(\chi|E|^2)E \tag{30}$$

where χ is the dielectric susceptibility

$$\chi = \alpha(1 - i\Delta)\left(1 + \Delta^2 + \frac{|E|^2}{I_s}\right)^{-1} \tag{31}$$

which has the form

$$\chi = \chi_a + i\chi_d \tag{32}$$

where χ_a and χ_d are the absorptive and dispersive parts of the susceptibility. The parameter Δ is the detuning between the incident light and the atomic transition frequency measured in units of γ,

$$\Delta = \frac{\omega_0 - \omega}{\gamma_\perp} \tag{33}$$

I_S is the saturation intensity given by

$$I_S = \frac{\hbar^2 \gamma_\perp \gamma_\parallel}{4|\nu|^2} \tag{34}$$

and α is the unsaturated absorption coefficient

$$\alpha = \frac{\omega_0 |\nu|^2}{2\epsilon_0 \hbar V c \gamma_\perp} \tag{35}$$

We first consider the case where the incident electric field, the atomic transition frequency, and the cavity are in perfect resonance; i.e., $\omega = \omega_0 = \omega_c$. This is the case of optical bistability that is purely absorptive. We define a normalized electric field F as

$$F = \frac{E}{\sqrt{I_S}} \tag{36}$$

and normalized incident and transmitted amplitudes x and y by

$$x = \frac{E_T}{\sqrt{I_S T}} \tag{37}$$

$$y = \frac{E_I}{\sqrt{I_S T}} \tag{38}$$

From Eqs. 30 and 31 we then have

$$\frac{dF}{dz} = -\alpha \frac{F}{1 + F^2} \tag{39}$$

and the boundary conditions become, at steady state,

$$x = F(L) \qquad F(0) = Ty + Rx \tag{40}$$

The differential equation in $F(z)$ can now be solved:

$$\ln\left(\frac{F(0)}{x}\right) + \frac{1}{2}[F^2(0) - x^2] = \alpha L \tag{41}$$

And combining Eqs. 37, 38, and 40 we can obtain the dependency of αL on x and y:

$$\ln\left[1 + T\left(\frac{y}{x} - 1\right)\right] + \frac{x^2}{2}\left\{\left[1 + T\left(\frac{y}{x} - 1\right)\right]^2 - 1\right\} = \alpha L \tag{42}$$

We now have three equations, Eqs. 40–41, which are adequate to describe the optical bistability of the model. The first two equations are both functions of the incident field intensity at the first mirror represented by $F(0)$ and the field intensity at $z = L$ represented by the parameter x (= $F(L)$). The first equation (40) is the boundary condition at the mirror. The second (41) is the transfer function for the field intensity within the medium, relating the field $F(0)$ to $F(L)$. If $R = 0$, then $F(0) = y$, and x and y are proportional to the incident and transmitted fields E_T and E_I, respectively. For other values of R the relation between x and y depends on the intersection between Eqs. 40 and 41.

The bistable condition can be observed by plotting $F(0)$ versus x for both Eqs. 40 and 41. This is shown in Fig. 8 [2]. Equation 40 is plotted for a value of $R = R_c$, where R_c is the tangential slope of the curve of Eq. 41 at its inflection point. This slope is such that $0 < R_c < 1$. For values of the reflectivity R such that $R \leq R_c$, there is only one intersection point for all values of y and one unique value of $F(0)$ for each x. For the condition $R > R_c$, however, the line represented by Eq. 40 intersects Eq. 41 at more than one point, and there is a range in values of y for which there may be three intersection points x_a, x_b, and x_c. Points x_a and x_c

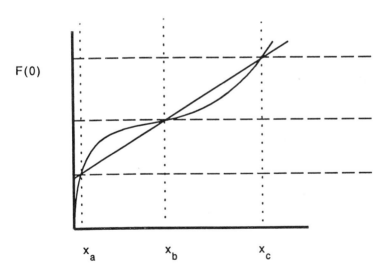

Figure 8   Graph of the normalized field intensity at $z = 0$ as a function of the normalized field intensity at $z = L$ ($F(0)$ vs. $x = F(L)$) Equation 40 in the text is plotted for a value of reflectivity $R = R_c$, where R_c is the tangential slope of the inflection point of Eq. 41. (From Ref. 2.)

turn out to be stable while x_b is unstable. Thus we have a bistable condition represented by two different normalized output field intensities, x_a and x_c, for a given normalized input field intensity y. The conditions on the value of R illustrate the importance of feedback in order to obtain bistability.

The S shape of the transfer curve can be observed by plotting x as a function of y (Eq. 42), as is done in Fig. 9 [2]. The figure is plotted as a function of $C = \alpha L/2T$ for a fixed value of C. As αL and T both decrease, the curve becomes S-shaped. This bistability is manifested by the combination of the nonlinear energy transfer of the medium, combined with the feedback from the mirrors. Both feedback and nonlinear susceptibility are required for bistability.

To produce purely absorptive bistability, the incident laser line can be tuned to the resonance of the medium. The threshold for bistability then depends upon the finesse of the cavity and the inhomogeneity of the absorption. In dispersive bistability, a nonlinear medium might include an electro-optic crystal. The incident light is monitored, and corresponding changes in the refractive index of the medium can then be produced.

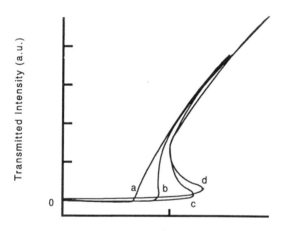

Figure 9 Graph of the normalized transmitted light intensity vs. incident intensity for a fixed value of $C = \alpha L/2T$. The transmissivity T was varied with αL in the figure as (a) $\alpha L = 20$, $T = 1$ (monostability); (b) $\alpha L = 10$, $T = 0.5$; (c) $\alpha L = 2$, $T = 0.1$; (d) mean field for the case $\alpha L \to 0$. (From Ref. 2.)

2.3 Mechanisms of Optical Nonlinearities in Semiconductors

Semiconductors as Etalons

An ideal material for practical optical bistable applications is one which has a large nonlinearity and a relatively low absorption at the wavelength of interest, as this is the optimum condition for dispersive effects. The larger the dispersive nonlinearities, the lower the optical power necessary to induce bistability. A fundamental problem is the known relation between absorption and dispersion.

As an example, in semiconductors most changes in refractive index from optical interactions are a result of a change in the free-carrier density. Refractive index changes can be induced by the formation of electron-hole pairs using incident light of energy equal to or greater than band-gap energy. These electron-hole pairs are, however, created by optical absorption. A change in dispersion is accompanied by a change in the absorption, and absorption represents a loss in dispersive bistable devices.

What is required is a situation similar to the two-level model described in Section 2.1, a saturation that can be achieved without absorptive loss. For most semiconductors at elevated temperatures, the use of intrinsic (i.e., larger than band-gap energy) light would not normally induce bistability as the absorption is broad and the absorption coefficient is large: the incident power to achieve saturation is prohibitively large.

An alternative to intrinsic light energy would be the use of light of energy "near band-gap," to induce exciton absorption at sub-band-gap light energies. The exciton is formed from the Coulombic attraction of the electron-hole pair. Compound semiconductors such as GaAs are known to possess a sharp atomic-like exciton resonance [3] with low saturation absorption intensities, particularly at low temperatures and high sample purity. Utilizing the excitonic transition makes the bistable etalon in a material such as GaAs more dispersive, allowing for fast switch times.

Usually the etalon is prepared in thin film or in crystalline plates. Epitaxial GaAs can be prepared in a high-purity form. Blocking layers can be grown from AlGaAs to help keep the GaAs excitons from recombining at the surface [4]. The thickness of the films can be adjusted to match a certain range of wavelengths from a tunable laser. The feedback is provided by the natural reflectivity of the semiconductor surface or by providing a reflective coating or a series of coatings.

In a linear material it is assumed that the dielectric susceptibility χ and the magnetic permeability κ are independent of the applied fields, and that the dielectric polarization P and magnetism M are both linear functions of these fields:

$$P = \chi E; \qquad M = (\kappa - I)H \tag{43}$$

The propagation of electromagnetic radiation in a material can be described by the wave equation:

$$\frac{\partial^2 E}{\partial z^2} - \frac{1}{c^2}\frac{\partial^2 E}{\partial t^2} = \frac{4\pi}{c^2}\frac{\partial^2 P}{\partial t^2} \tag{44}$$

The relationship between the dielectric constant and the dielectric susceptibility can be found from the D field:

$$D = E + 4\pi P = \epsilon E = (1 + 4\pi\chi)E \tag{45}$$

where the linear polarization

$$P = \chi E \tag{46}$$

has been used. This derives the relationship between susceptibility and the dielectric constant:

$$\epsilon = 1 + 4\pi\chi \tag{47}$$

In a nonlinear optical material, the polarization is often expanded as a power series over the electric field:

$$P_i = \chi_{ij}^{(1)}E_j + \chi_{ijk}^{(2)}E_{jk} + \chi_{ijkl}^{(3)}E_{jkl} + \cdots \tag{48}$$

where the subscripts refer to the Cartesian coordinate system. The higher orders of χ are called the nonlinear susceptibilities and can be high-ranking tensors depending upon the crystal symmetry. In this case we have assumed the polarization only depends upon the instantaneous value of the field amplitudes. It is particularly appropriate as the nonlinearities are often very small compared to the first-order (linear) susceptibility.

The first term, $\chi^{(1)}$, on the right-hand-side of Eq. 48 describes the linear optical properties of the system. Those terms following are called the nonlinear suscepibilities. The second-order term $\chi^{(2)}$ describes second-harmonic sum and difference effects, while $\chi^{(3)}$ describes third-order nonlinear effects. Such third-order effects include coherent anti-Stokes Raman scattering (CARS) and four-wave mixing as discussed in Section 4.

In linear optics the optical properties of a material such as transmission, reflection, refraction, and the complex dielectric function

$$\epsilon = n + ki \tag{49}$$

where n is the refractive index and k is the extinction coefficient, are all dependent on the frequency of the incident light. In birefringent materials, *n* and *k* depend on the electric field polarization as well as the propagation direction. In nonlinear materials these properties are also dependent on the light intensity and, thus, the strength of the electric field amplitude. A material is termed linear or nonlinear depending on whether the polarization of the electric dipoles produced by the electric field of the light is linear or nonlinear.

There is a large number of possible electronic excitations in a semiconductor, some of which form various microscopic entities such as excitons. Because of their large number and the many possible electronic interactions of these excitations, calculations describing nonlinear optical effects in semiconductors involve many-body problems rather than the more simplified single-particle model usually used to describe nonlinear phenomena.

There are various different microscopic mechanisms of optical nonlinearities in semiconductors. In a real system these mechanisms sum together to describe the polarization contribution to the macroscopic susceptibility and usually cannot be separated. Many physical explanations center only on the nonlinear susceptibility and are not concerned with the deeper causes of nonlinearities.

Physical mechanisms of nonlinearities can normally be classified into those involving the generation of free carriers (sometimes called optoelectronic) and those involving thermal effects (optothermal). Optoelectronic nonlinearities are the most common. These involve the optical excitation of electrons and holes between energy states. These excitation mechanisms can be further subdivided into virtual and real excitations.

During real excitation, incident radiation of energy above the band-gap energy excites carriers in a single-photon absorption process. The result is the creation of an electron-hole pair across the band-gap, the Coulombic attraction of which can bind the electron and hole together to form an exciton.

Virtual transitions are formed using photons of energy less than band-gap energy, which is below the fundamental absorption edge of the semiconductor. In a virtual optical transition, the coherent electric field of a laser induces a polarization in the material, the strength of which is dependent on the intensity of the exciting electric field. The electric field strength of the optical field $E(\omega)$ is increased until it becomes comparable to the atomic fields of the atoms holding the valence electrons. The optical field induces an anharmonic charge density fluctuation within the atom which, in turn, can interact with other optical fields. This is done without depositing any energy into the material (hence a virtual transition). The two optical fields coupled by the charge fluctuation can now exchange photons between them. The energy states of the atom are coherently coupled by the two photons and the atom acts as if were excited, although no actual net exchange of carriers between energy states exists. The result is a two-photon transition to an exciton state, with an exciton (the virtually excited atom) being the intermediary for the transition. These virtual transitions result in a nonlinear, intensity-dependent susceptibility.

The optical nonlinearities involved in virtual two-photon transitions are weak due to the low population density of the virtually excited states, but have fast response times (on the order of $\Delta E/\hbar$, where ΔE is the difference in energy between the photon and the excited states).

It is difficult experimentally to have all virtual transitions present in an excited semiconductor, without having an appreciable quantity of real optical transitions, as well as some thermally stimulated transitions. In an actual semiconductor the nonlinearities produced are primarily generated from electron-hole pairs (free carriers or bound as excitons) generated in real excitation processes. In a practical optical bistable system suitable for making optical devices, bistability is produced from free carriers because band tails of the density of states exist extending into the band gap (see Section 10 in Chapter 1). At high enough power densities the sub-band-gap energy does create a finite concentration of electron-hole pairs in addition to the virtual transitions.

In narrow-gap compounds such as InSb, the optical nonlinearities are caused by band filling, while in larger-gap materials, such as GaAs and CdS, excitonic effects are more important. The optical absorption edge of InSb lies near 7 μm at room temperature and near 5.5 μm both at liquid nitrogen and at liquid helium temperatures. This is a convenient material to experiment with, as a CO_2 laser has several lines in the 5–6-μm region. Excitation of InSb with light of this spectra will induce nonlinearities from a reduction of the refractive index due to the saturation of the band absorption, just as the saturation of the absorption leads to saturation of the refractive index in the two-level system. The resulting optical susceptibility becomes dependent upon the density of the excited carriers.

Excitons have energies slightly below band-gap energy and are most attractive for the generation of optical nonlinearities. The resonant absorption lines of excitons are broadened at ordinary temperatures by electron-phonon interactions. If light of high enough intensity from a laser source of subgap energy is incident on the sample, enough excitons can be generated at one time that the separation between neighboring excitons becomes comparable to the individual Bohr radius of that exciton, and then exciton-exciton interactions become more important than electron-hole-phonon interactions in determining the absorption lineshape. In so-called highly excited semiconductors, exciton-exciton scattering processes have been identified, along with the formation of new entities such as exciton molecules [5].

As the light intensity increases, the concentration of these electron-hole pairs continues to increase, the screening of the Coulombic forces increases, and the Coulombic attraction between electron-hole pairs becomes weaker. In a bulk semiconductor such as GaAs, at a critical light intensity the screened Coulombic potential is too weak to support a bound state and the excitons cease to exist. The excited electrons and holes then form a two-component plasma. The semiconductor has made a transition from an insulator to a material with metallic-like properties, and the two-component plasma then influences the optical properties of this material.

In an indirect-gap material the electron-hole lifetime is long enough that a stable "plasma liquid" can be formed at low temperatures and high enough laser intensities. The population of these excited states is usually too small, however, for use in a practical device, and the optical nonlinearities are quite weak.

In a direct-gap semiconductor the existence of the electron-hole plasma exerts large changes of the optical properties near the absorption edge. The semiconductor makes a transition similar to a Mott transition from a semiconductor to a metal. The sharp exciton resonance lines disappear, and the material becomes increasingly opaque as the large number of photoexcited electrons and holes leads to a reduction in the band-gap energy (band-gap renormalization). The physics of plasma gases in crystals is not simple, and predictions of optical properties are difficult to make.

As a consequence of the band-gap reduction with increased light intensity, the band-gap energy then is reduced to the energy of the incident radiation, and there is an increase in the optical absorption.

Thermal Optical Bistability

A good portion of the observed optical nonlinearities arise from localized heating from the optical energy. Excited electron-hole pairs can recombine in both radiative and nonradiative mechanisms. Nonradiative recombination is accompanied by a transfer of thermal energy to the lattice. This heats the lattice, expanding the spacing between atomic planes, at the same time shrinking the band gap. The absorption thus increases, increasing the electron-hole pair density, the recombination rate, and thus the lattice temperature. A nonlinear absorption which increases with intensity is the result. In this case, the absorption is a function of the lattice temperature in the same way that it was a function of the carrier density N in a purely optoelectronic nonlinear absorption. That is,

$$\alpha = \alpha(\omega, T) \tag{50}$$

Photothermal nonlinearities are favorable at room temperature and for long excitation pulses.

Equation 51 [6] describes the relationship between the change in refractive index with temperature and the band-gap energies for near-band-edge excitations:

$$\frac{dn}{dT} = E_g^{-1/2} \left(\frac{E_g}{E_g - \hbar\omega} \right)^{1/2} \tag{51}$$

The equation illustrates how the variation is greatest for excitation closest to band-gap energies.

Optical Bistability without an Etalon

Optical bistability can be achieved without the use of mirrors as a source of feedback. In this case, absorption increases with increasing light intensity to produce optical bistability. If one has a set of absorption centers N, where the absorption is a function of the value of N and the intensity I,

$$\alpha = \alpha(N, I) \tag{52}$$

and the value of N, in turn, is dependent upon the intensity of the light,

$$N = N(I, \alpha) \tag{53}$$

then, N also depends on the absorption. Equations 52 and 53 are two identities which may give more than one solution of α for a given I, and hence the system is bistable. This is often termed bistability without feedback, but if one looks closely one can observe that the value of N is a feedback parameter. Increasing the light intensity increases N, which increases the absorption, which increases N further, which increases the absorption further, etc.

We have already seen how optical nonlinearities can be induced in semiconductors through band-gap shrinkage either by excitation of intensity-dependent resonances or through thermal shifts of those resonances. In the case of intensity-dependent concentrations, the value of N is coupled to the intensity through rate equations such as [7]

$$\frac{\partial N}{\partial t} = \frac{\alpha I}{\hbar\omega} - \frac{N}{\tau} + \frac{\partial D}{\partial z} \frac{\partial N}{\partial z} \tag{54}$$

where the beam is propagating in the z-direction: D is the electron-hole pair diffusion coefficient and τ is the carrier lifetime. A simulated plot of α versus N is shown in Fig. 10. The dashed line represents Eq. 54, while the solid line represents a relation between the density and intensity of the light such as

$$N = \alpha\eta I \tag{55}$$

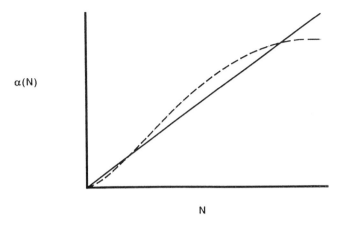

$\alpha(N)$

N

Figure 10 Density dependence of absorption (dashed line) and relation between density and illumination intensity (solid line) for a semiconductor such as CdS. Multiple intercepts of the two curves show how two separate outputs can exist for the same input and hence be bistable without the need of a resonator cavity.

where we have assumed linear dependence and η is the constant of proportionality.

As can be observed from the figure, there is a possibility of having more than one intersection of the two curves, and thus two possible output states for this one input state. Increase of the incident intensity from a low to a high level satisfies Eq. 54 at a different point than it does when decreasing from a high to lower level, thus producing a hysteresis curve.

Multi-Quantum-Well Devices

An interesting development in the study of multi-quantum wells is the quantum-confined Stark effect. In direct-gap semiconductors, the application of an electric field to a device will normally rip apart an electron and hole and destroy the exciton. In a MQW the walls of the potential well physically prevent separation of the charge, and the Coulombic attraction between the carriers remains quite high within the well, even under an electric field.

The Stark effect in semiconductors can be reviewed in Chapter 1. The applied electric field first polarizes the charge cloud of the exciton. As the electron and hole align themselves with the electric field, and the orbit of the exciton becomes increasingly elliptical, the energy of the exciton becomes subsequently smaller, and the energy difference between the exciton and the band-gap energy increases. This shifts the exciton absorp-

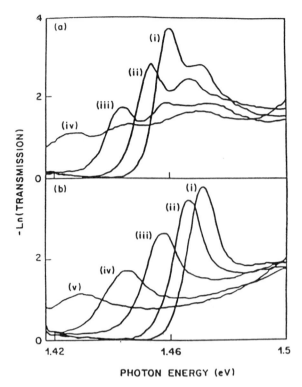

Figure 11 Absorption spectra of an optical waveguide fabricated from multi-quantum wells with an electric field applied perpendicular to the layers. (a, b) Incident optical polarization field perpendicular (a) and parallel (b) to the plane of the layers under electric fields of (a)(i) 1.6×10^4 V/cm, (ii) 10^5 V/cm, (iii) 1.3×10^5 V/cm, (iv) 1.8×10^5 V/cm. (b)(i) 1.6×10^4 V/cm, (ii) 10^5 V/cm, (iii) 1.4×10^5 V/cm, (iv) 1.8×10^5 V/cm, (v) 2.2×10^5 V/cm. (From Ref. 8.)

tion resonances away from the band edge to lower energies. Figure 11 [8] shows the experimentally measured shift of exciton resonances with electric field. Because the exciton resonances are so sharp, as the resonances move to lower energies, the absorption at any one photon energy decreases. Hence, the optical absorption at the initial photon energy (the so-called zero field energy) decreases with increasing electric field. Separating exciton resonances from the absorption means the nonlinearities will be more dispersive with fewer absorption losses.

If the size of the quantum-confined Stark effect can be maximized, the exciton resonances will be observed at higher temperatures. A near-room-temperature MQW exciton device has been theorized.

3 SELF-ELECTRO-OPTIC EFFECT DEVICE (SEED)

The self-electro-optic effect device is an optical bistable device which has an electronic feedback component. The SEED is usually fabricated in multi-quantum-well structures, taking advantage of the sharp near-room-temperature exciton resonances observed in those structures.

As noted, excitons in direct-gap semiconductors are usually only observed as sharp resonances at low temperatures. This is because at elevated temperatures, the resonances broaden due to interactions with thermal phonons. However, confining the electrons in a multi-quantum-well structure (see Chapter 1, Section 14) in a dimension less than the exciton Bohr diameter (about 300 Å in GaAs) increases the separation of the exciton resonances from the band-gap absorption (the so-called bound exciton energy) without increasing the phonon broadening. Moving the exciton resonances away from the absorption edge means nonlinear devices can be tuned to the resonances with fewer absorption losses.

The SEED can be thought of as a combination of an electro-optic modulator and a semiconductor photodetector. Figure 12 shows a schematic of such a device. The photodetector, on sensing light, produces a photocurrent. This photocurrent influences the voltage across the modulator, which is simply the electric field across the MQW structure. The presence of the photocurrent reduces the electric field across the device, and the absorption of photons of energy corresponding to the zero-field exciton resonance increases. As the absorption increases, the photocurrent increases, and the resonance continues to shift in a direction that further

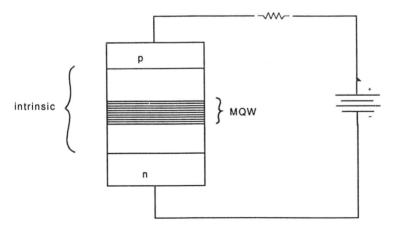

Figure 12 Schematic of a self-electro-optic effect device based on multi-quantum well structures.

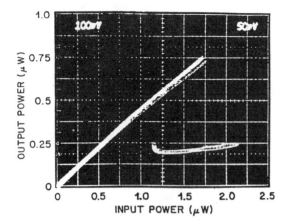

Figure 13 Superimposed input-output characteristics of four devices in an integrated SEED array. (From Ref. 8.)

increases the absorption. The result is a positive feedback system, with the electrical current from the detector acting as the feedback mechanism. The positive feedback causes the SEED to switch into a high-absorption state. Figure 13 [8] shows the input-output characteristics of four devices superimposed.

4 PHOTOREFRACTION

Photorefraction is a name given to a third-order nonlinear optical effect observed in certain insulators and semi-insulators. Photorefraction is a method of writing volume holograms which are dynamic, i.e., volume gratings which erase after cessation of the writing beams with time constants which vary from milliseconds to several months. Thus it holds promise for many of the suitable applications of nonlinear mixing, including spatial filters, phase conjugation using degenerate four-wave mixing, optical processing devices such as optical oscillators, and as a technique useful for nondestructive, noncontacting materials analysis.

Photorefraction has been observed in several insulators, mostly ferroelectrics with large electro-optic coefficients and small dielectric constants. An optimum photorefractive crystal must

1. Be fairly transparent at the wavelength of interest
2. Have a relatively large concentration of deep defect centers in its band gap

3. Possess a large electro-optic coefficient/dielectric constant ratio
4. Be a relatively good photoconductor

A simplified photorefraction model is the single-carrier model summarized mathematically by Kukhtarev [9]. Two coherent pump beams enter a photorefractive crystal with an angle of 2Θ between them, as shown in Fig. 14. The Kukhtarev model of the intersection point of the two beams is shown in Fig. 15. Where the coherent pump beams mix they form an intensity interference pattern of alternating light and dark areas that can be modeled as

$$I(x) = I_0 + I_1 e^{i(\Omega t - Kx)} + \text{c.c.} \tag{56}$$

where c.c. represents the complex conjugate of the first argument of the equation.

At the high-intensity regions of the interference pattern, the light is absorbed in the crystal through altering of the population of deep traps in the band gap. In the single-carrier model the light is absorbed by exciting electrons (holes) to conduction band (or valence band). The free carriers diffuse (or drift if an external electric field is applied) to dark areas of the interference pattern where they recombine. The result is a periodic space-charge region, as shown in Fig. 15b. This space-charge region results in a periodic electric field, the amplitude of which (with no external field) is

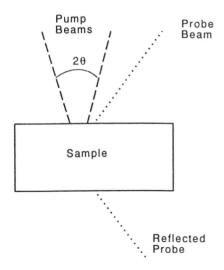

Figure 14 Schematic of the nondegenerate four-wave mixing experiment involved in observing the photorefractive effect.

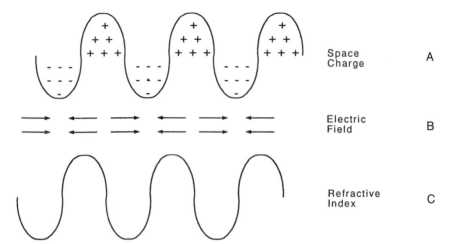

Figure 15   Single-carrier model illustrating the photorefractive effect. An interference pattern (A) creates a modulated space-charge region (B) by way of the photoconductive effect, which causes a refractive index grating to form (C) by way of the linear electro-optic effect.

determined as

$$E_{SC} = -i \frac{I_1}{I_0} \frac{E_0}{1 + E_0/E_N} \tag{57}$$

where

$$E_N = \frac{eN_A}{\epsilon K}, \qquad E_0 = \frac{k_B T K}{e} \tag{58}$$

The significance of Eq. 57 being imaginary is a 90° phase shift between the field and the interference pattern. Since the material is also electro-optic, the refractive index is a function of the electric field (see Chapter 8). The electric field causes a periodic change in the refractive index from the third-order electro-optic coefficient defined from

$$\Delta n \propto r E_{SC} \tag{59}$$

The resulting refractive index modulation is modeled as

$$n(x) = n_0 + n_1 e^{i(\Omega t - Kx)} + \text{c.c.} \tag{60}$$

The result of the mixing of the two pump beams of wavelength λ is a refractive index grating having grating spacing given by Bragg's law:

$$\lambda = 2\Lambda \sin \theta \qquad (61)$$

A grating vector will be defined in space and time by the relative magnitude and direction of K and Ω.

The result is a refractive index grating of spacing $\Lambda = 2\pi/K$ from which a third beam of light (referred to as the "probe") can diffract in a similar way to which a beam of light may diffract from a fixed grating etched in glass. If the probe beam (dashed line shown in Fig. 14) is of a wavelength larger than the pump beam, it then diffracts at a Bragg angle larger than the pump beam as demanded by Eq. 61. This is the so-called nondegenerate four-wave mixing configuration. The probe beam produces a diffracted fourth beam which can be used as a means of monitoring the grating.

In so-called fully degenerate four-wave mixing, a probe beam of the same frequency and polarization as the pump is used. Then a fourth beam is diffracted at angle Θ_{pump} backwards along the same direction as one of the pump beams. Due to the nonlinear nature of the mixing in the crystal this fourth beam is the spatial complex conjugate of the pump beam it follows along. Due to the possibility of beam coupling, one pump beam is diffracted into the other; a reflection of one of the coupled beams back into the crystal thus can be used as "self-pumped" phase conjugation—that is, phase conjugation using only one beam from one laser.

In two-beam coupling, instead of using a probe beam one pump beam self-diffracts from the grating. The significance of Eq. 57 being imaginary is a quarter-period phase shift between the electric field and the light intensity. This phase shift can be observed by comparing Figs. 15a–c. This quarter-period shift allows one pump beam to self-diffract from the grating into the other beam. Thus, a power transfer between the two pump beams exists and is referred to as two-beam coupling. Beam coupling can be observed without the use of the probe beam. For some ferroelectrics the magnitude of energy transfer in two-beam coupling can approach 100%; one pump beam emerges from the crystal almost extinguished, the other at almost twice its incident intensity.

Photorefraction is unique in that the nonlinear effect does not require high-intensity sources. This is because the grating modulation index m does not depend on the absolute intensity of the pump beams but on the ratio of pump beam intensities. Thus, it is conceivable that in the right material it could be observed using available low-power coherent sources. The speed of the effect does, however, depend on light intensity.

The time dependence of the index grating is determined by [10]

1. Dielectric relaxation time
2. Diffusion length of the photocarriers
3. Debye screening length

Which of these, or which combination of these, parameters is important is dependent upon the intensity of the light and the grating period. When making a grating, the rise time of the diffraction efficiency increases if total writing intensity increases. The writing and erasing rates increase if the angle between the pump beams increases [10]. Erasing times can also be increased with an erase beam, which offers a very short relaxation time by altering the photoconductivity of the sample.

REFERENCES

1. H. M. Gibbs, S. L. McCall, and T. N. C. Venkatesan, Differential gain and bistability using a sodium-filled Fabry-Perot interferometer, *Phys. Rev. Lett.*, *36*, 1135 (1976).
2. L. A. Lugiato, Theory of optical bistability, in *Progress in Optics*, Vol. XXI (E. Wolf, ed.), North-Holland, Amsterdam, 1984. p71.
3. D. D. Sell, S. E. Stokowski, R. Dingle, and J. V. DiLorenzo, Polariton reflectance and photoluminescence in high-purity GaAs, *Phys. Rev. B*, *7*, 4568 (1973).
4. H. M. Gibbs, A. C. Gossard, S. L. McCall, A. Passner, W. Wiegmann, and T. N. C. Venkatesan, Saturation of the free exciton resonance in GaAs, *Solid State Commun.*, *30*, 271 (1979).
5. S. W. Koch, Optical instabilities in semiconductors: theory, in *Optical Nonlinearities and Instabilities in Semiconductors* (H. Haug, ed.), Academic, 1988, p. 273.
6. B. S. Wherrett, A. C. Walker, and F. A. P. Tooley, Nonlinear refraction for CW optical bistability, in *Optical Nonlinearities and Instabilities in Semiconductors* (H. Haug, ed.), Academic, 1988, p. 239.
7. Y. Toyozawa, Bistability and anomalies in absorption and resonance scattering of intense light, *Solid State Commun.*, *28*, 533 (1978); *32*, 13 (1979).
8. D. A. B. Miller, D. S. Chemla, and S. Schmitt-Rink, Electric field dependence of optical properties of semiconductor quantum wells: physics and applications, in *Optical Nonlinearities and Instabilities in Semiconductors* (H. Haug, ed.), Academic, 1988, p. 325.
9. N. V. Kukhtarev, V. B. Markov, S. G. Odulov, M. V. Soskin, and V. L. Vinteskii, Holographic storage in electrooptic crystals. I: Steady state," *Ferroelectrics*, *22*, 949 (1979).
10. G. C. Valley and P. Yeh (Guest Eds.), *J. Opt. Soc. Am. B Special Issue on Photorefractive Effects and Devices*, *5*, 1682 (1988), and references therein.

7

Acousto-Optic Devices

1 INTRODUCTION

When a sound wave passes through a material, it produces a localized strain field within the medium. This strain field consists of areas of tension and compression which are dynamic; the strain field travels with the acoustic wave as it passes through the material. The velocity of this strain field is dependent upon the medium in which it is propagating. By way of the acousto-optic effect, this localized tension and compression of the material produce variations in the refractive index able to alter the propagation of light through the medium. By adjusting the acoustic wave, one can observe an affect on the optical wave.

The acousto-optic device is found in many applications, including optical modulation and deflection, Q-switching, and mode locking. In communications, acousto-optic devices are used for frequency shifting and channel switching, multiplexing, and demultiplexing. In RF signal processing, a-o devices can be found in use for convolution, correlation, and spectrum analysis. Finally, there is an ongoing effort to use these devices in optical processing.

Figure 1 illustrates an example of how sound can alter the propagation of an optical wave. An acoustic wave or beam is sent into a material (normally a good piezoelectric). The material is also optically transparent, and an optical beam incident at an oblique angle to the direction of the

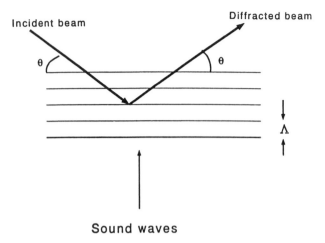

Figure 1 The acousto-optic effect as realized in Bragg diffraction. An incident optical beam is reflected off a refractive index grating caused by the stresses built into a material by a traveling acoustic wave.

propagation of the acoustic wave is diffracted from the material. This beam would normally pass through the material unaffected in the absence of the acoustic wave.

In the presence of the acoustic wave, the medium becomes optically inhomogeneous. The inhomogeneity exists as a refractive index grating that is dynamic; that is, the grating also propagates as the acoustic wave propagates through the crystal. This dynamic perturbation of the materials index propagates at a velocity slow compared to the velocity of the optical beam. While there is a small "Doppler shift" of the reflected light in Fig. 1 due to the velocity of the acoustic wave, it can, under normal analysis, be ignored. The crystal can then be analyzed as a static refractive index grating. The acousto-optic effect can then be modeled as periodic static gratings. For the bulk acousto-optic effect, diffraction as illustrated in Fig. 1 is known as the *Bragg* condition of diffraction.

2 BRAGG REFLECTION

In a bulk acousto-optic crystal, the diffraction of light off of a periodic refractive index grating can be modeled as simple reflection, as illustrated in Fig. 1. An acoustic wave traveling along a direction x through the crystal sets up a periodic quasi-static strain field in the crystal modeled as

$$\epsilon(x, t) = \epsilon_0 \cos(Kx - \Omega t) \tag{1}$$

where ϵ_0 is the amplitude of the strain and Ω and K are the temporal and spatial frequencies (radial) of the acoustic wave. The index grating spacing is given by the wavelength, Λ, of the acoustic wave:

$$\Lambda = \frac{2\pi}{K} \tag{2}$$

The strain field creates a variation in the refractive index determined by

$$\Delta \left(\frac{1}{n^2}\right)_{id} = p_{idkl}\epsilon_{kl} \tag{3}$$

where the subscripts *ikl* refer to various directions in the crystal and p_{idkl} is the *photoelastic tensor*. Application of stress field σ_{kl} causes formation of an index ellipsoid (see Chapter 8) of index constant given by $\Delta(1/n^2)_{id}$. If the acoustic wave is applied along a preferential direction of the crystal, and that direction is the coordinate x, the subscripts in Eq. 3 can be dropped and the equation can be written as

$$\Delta n(x, t) = \frac{-1}{2} pn^3 \, \epsilon(x, t) \tag{4}$$

where the stress wave produces a variation in strain in x and t. A positive strain reduces the refractive index. The acoustic intensity, in W/m², is given by

$$I_a = \frac{1}{2} \rho v^3 \epsilon_0^2 \tag{5}$$

where ρ is the density of the material and v is the velocity of the acoustic wave. The dispersion is assumed to be linear with $v = \Omega/K$. The refractive index can be written as a space-time-variable wave:

$$n(x, t) = n_0 - \Delta n_0 \cos(Kx - \Omega t) \tag{6}$$

where n_0 is the refractive index in the absence of the strain field and Δn_0 is the amplitude of the strain. The amplitude is

$$\Delta n_0 = \frac{1}{2} pn^3 \epsilon_0 = \left(\frac{1}{2} M I_a\right)^{1/2} \tag{7}$$

where the material parameter

$$M = \frac{p^2 n^6}{\rho v^3} \tag{8}$$

To consider the interaction of the optical beam with this refractive index grating, we consider several planes of refractive index as shown in Fig. 2. The layers are spaced a distance Λ from one another and the entire structure is of thickness L, which represents the penetration distance of the acoustic wave in the x direction. As the acoustic wave velocity is slow compared with the light velocity, the grating will be treated as semistatic:

$$n(x) = n_0 - \Delta n_0 \cos(Kx - \phi) \tag{9}$$

where ϕ is now a fixed phase lag. An optical beam is incident upon the fixed grating at an angle θ with respect to the x direction, as shown in the figure. The reflectance of the optical beam can be analyzed by using the Fresnel equations for reflections from a medium of refractive index n into a medium of refractive index $n + \Delta n$. The individual change in reflectivity dr for a variation of refractive index dn can be shown as [1]

$$\frac{dr}{dn} = \frac{-1}{2n \sin^2 \theta} \tag{10}$$

In this case, Δr is the individual reflectance from a layer at position x in the stack where $\Delta r = (dr/dx)\Delta x$. For overall length L, the total reflectance r can be written as

$$r = \int_{-L/2}^{L/2} \exp[i2kx \sin \theta] \frac{dr}{dx} \, dx \tag{11}$$

The phase factor $\exp[i2kx \sin \theta]$ is included since only the $2x \sin(\theta)$ component is advanced between $x = 0$ and $x = L$. The wavevector k is that of both incident and reflected optical waves. We can write

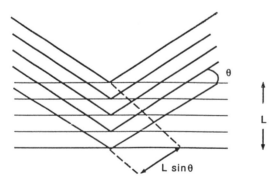

Figure 2 The reflection of an optical beam from an a refractive index grating of grating period Λ and total depth L.

$$\frac{dr}{dx} = \frac{dr}{dn}\frac{dn}{dx} = \frac{-K\,\Delta n_0\,\sin(Kx - \phi)}{2n\,\sin^2\theta} \tag{12}$$

Substituting Eq. 12 into Eq. 11 and writing $\sin(Kx + \phi)$ as a complex exponential yields

$$r = \frac{1}{2}\,ir'e^{i\phi}\int_{-L/2}^{L/2} e^{i(2k\,\sin\theta - K)x}\,dx - \frac{1}{2}\,ir'e^{-i\phi}\int_{-L/2}^{L/2} e^{i(2k\,\sin\theta + K)x}\,dx \tag{13}$$

where

$$r' = \frac{-K\,\Delta n_0}{2n\,\sin^2\theta} \tag{14}$$

Integration of either the first or second term separately in Eq. 13 yields a sinc function for each term. The sinc function of the first term will dominate for $2k\,\sin(\theta)$ close to K, and the second term will dominate for $2k\,\sin(\theta)$ close to $-K$. Evaluating the integral in the first term for the condition $2k\,\sin(\theta) \approx K$ gives the sinc function

$$r = \frac{1}{2}\,ir'L\,\text{sinc}\!\left[(K - 2k\,\sin\theta)\frac{L}{2\pi}\right]e^{i\Omega t} \tag{15}$$

A plot of the reflectance given by $|r|^2$ is shown in Fig. 3. The sinc function is a maximum at $\theta = \theta_B$, where

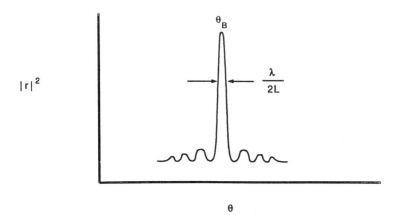

Figure 3 A plot of the reflectance $|r|^2$ vs. incident angle θ. The reflectance is a maximum for $\theta = \theta_B$, the Bragg condition.

$$\theta_B = \sin^{-1}\left(\frac{K}{2k}\right) \tag{16}$$

is the Bragg angle. This is the so-called Bragg condition of diffraction:

$$\sin \theta_B = \frac{\lambda}{2\Lambda} \tag{17}$$

The function drops off sharply on each side of the "peak" in Fig. 3. The function given by Eq. 15 is a minimum at $\sin \theta - \sin \theta_B = \lambda/2L$, where the sinc function reaches its first zero. For acousto-optic applications the angle of incidence is small, and $\sin \theta \approx \theta$. The narrow linewidth is defined by $\theta - \theta_B \approx \lambda/2L$. For large L this means the linewidth is very narrow, and the diffraction acts as a "slit," only passing light within a narrow cone of radiation. The Bragg diffraction acts as an optical-frequency-pass filter, as only light of a wavelength close to λ will diffract for a given θ and grating spacing Λ.

In the quantum-mechanical approach to Bragg diffraction, the optical beam interacts directly with the sound wave. An incident photon loses its energy, a new reflected photon is generated, and conservation of momentum dictates that $\mathbf{p}_r = \mathbf{p}_i + \hbar\mathbf{K}$, where $\mathbf{p}_r$, $\mathbf{p}_i$ are the momentum vectors of the reflected and incident optical waves, respectively, and $\hbar\mathbf{K}$ is the momentum vector of the lattice vibration. Quantized lattice vibrations are known as *phonons*. Since $\mathbf{p} = \hbar\mathbf{K}$, conservation of momentum yields

$$\hbar\mathbf{k}_r = \hbar\mathbf{k}_i \pm \hbar\mathbf{K} \tag{18}$$

which indicates that the Bragg condition can be written in terms of a vector sum. This is illustrated in Fig. 4. The choice of positive or negative in the r.h.s. of Eq. 18 depends on whether the phonon is adding or removing momentum from the process. In the majority of cases the photon loses energy to the phonon, and the sign is positive.

Likewise conservation of energy dictates

$$\hbar\omega_r = \hbar\omega_i + \hbar\Omega \tag{19}$$

where $\hbar\omega_i$ and $\hbar\omega_r$ are the incident and reflected photon energies, respectively, and $\hbar\Omega$ is the energy of the phonon. The energy $\hbar\Omega$ is the Doppler shift that the photon experiences. The frequency of the reflected light experiences a shift equal to the sound frequency. Since phonon energies are small compared to phonon energies, the Doppler shift is very small, and $\omega_r \approx \omega_i$.

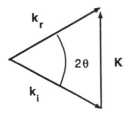

Figure 4 The Bragg condition as illustrated by vector sum $\mathbf{k}_r = \mathbf{k}_i + \mathbf{q}$, where $\mathbf{k}_i$ and $\mathbf{k}_r$ are the incident and reflected wavevectors, respectively, and $\mathbf{q}$ is the wavevector of the lattice vibration.

3 RAMAN-NATH DIFFRACTION [2]

If sound waves are sent through a very thin acousto-optic cell, the resulting refractive index grating will behave as a transmission grating. Such diffraction is inherently different from Bragg diffraction and is illustrated in Fig. 5. Diffraction from a Raman-Nath grating undergoes constructive interference whenever the diffracted angle θ is such that

$$\sin\frac{\theta}{2} = \frac{\Lambda}{2\lambda} \tag{20}$$

If θ is small, then

$$\theta = \frac{\Lambda}{\lambda} \tag{21}$$

and the thin acousto-optic cell acts as a diffraction grating of period Λ.

In the case of the bulk Bragg acousto-optic modulator, we assumed that both the acoustic and optical beams were plane, single-component collimated waves. In reality, both beams have a finite spectral width, and

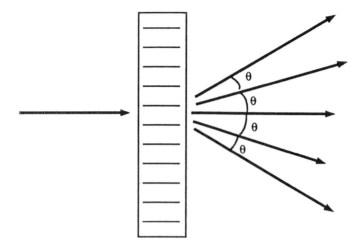

Figure 5   Physical arrangement for Raman-Nath diffraction.

each contains many components of plane waves. The wavelength of the sound beam is relatively large, the spectral range broad, and the optical component has a large number of acoustic plane waves with which to interact.

Figure 6 illustrates the diffracted components of a thin acousto-optic beam. The radiation pattern for the sound beam includes a central propa-gating wave of wavevector K, plus first- and second-order spectral compo-nents K_{-1} and K_{-2} as well as the symmetrical components K_{+1} and K_{+2}. Light is incident upon the cell perpendicular to the direction of travel of the zeroth-order acoustic vector (the so-called Debye-Sears case of Raman-Nath diffraction). Since the light is incident perpendicular to the zeroth order, the Bragg angle can never be satisfied for that order, and a portion of light k_0 (a zeroth-order light spectral component) passes through the light undiffracted.

Part of the incident light will also interact with other components of the acoustic wave. As shown in Fig. 7, part of the light will meet the Bragg condition with the K_{-1} order producing a downshifted optical com-ponent k_{-1}. This component will propagate at a $+\theta$ direction to the zeroth order. The symmetrical component about the vertical axis, K_{+1} will also produce Bragg diffraction with the incident angle, and an upshifted compo-nent k_{+1} is produced at angle $+\theta$. If the light is intense enough, second-order beams k_{+2} and k_{-2} can be produced by the interactions of k_{+1} with K_{+2} (producing k_{+2}) and k_{-1} with K_{-2} (producing k_{-2}, not shown in the

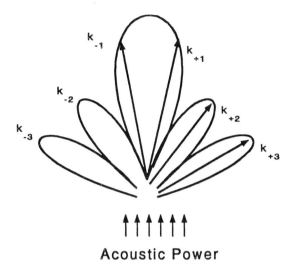

Acoustic Power

Figure 6 Diagram showing interaction of light and acoustic waves in multiple orders.

figure). If the strain wave is a standing wave, (which it virtually is, since the velocity of the acoustic wave is so small compared to the velocity of the optical wave), the incident light is diffracted both in the $+\theta$ and $-\theta$ directions. Figure 7 shows how optical waves that undergo multiple-order diffraction can be represented as a vector summation.

Quantum-mechanically speaking, both momentum and energy of the interaction must be conserved. The energy of the acoustic wave is quan-

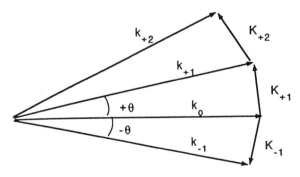

Figure 7 Conservation of momentum diagram of the process in Fig. 6.

tized, and has value $\hbar\Omega$ while the momentum is given as $\hbar\mathbf{K}$. Momentum and energy are transferred to and from incident photons, the interaction thus producing exiting photons.

For example, for a second-order reflection, one incident photon of momentum $\mathbf{k}$ must interact with two phonons to form the second-order reflection

$$\mathbf{k}_0 = \mathbf{k}_2 + 2\mathbf{K} \tag{22}$$

Diffraction theory demonstrates that for an acoustic beam of thickness L, the width of the acoustic spatial spectrum is proportional to Λ/L, where Λ is the wavelength of the acoustic signal. The Bragg condition for narrow incident angles tells us that the orders are separated by the angles $\pm\theta \approx \pm\lambda/\Lambda$. Dividing the angular separation of the orders into the angular width will give the number of possible orders that may appear. The inverse of this number over 2π is called Q and is used to determine the condition for Raman-Nath diffraction:

$$Q = \frac{2\pi\lambda L}{\Lambda^2} \tag{23}$$

For $Q \gg 1$ the system is said to be in the Bragg regime; for $Q \ll 1$ it is in the Raman-Nath regime.

4 ACOUSTO-OPTIC MODULATORS, DEFLECTORS, AND SWITCHES [3]

An acousto-optic device operating in the Bragg mode can perform the role of an optical switch, modulator, or deflector simply through control of the RF signal feeding the transducer. Such devices may be prepared as bulk components or as thin-film waveguides in $LiNbO_3$ or GaAs and are capable of operation in the GHz range.

For a sufficiently strong RF signal, and thus strong acoustic signal, a Bragg acousto-optic device can perform the function of a beam deflector or steerer, deflecting the incident optical beam into the 2θ direction, where

$$\theta = \sin^{-1}\frac{\lambda}{2\Lambda} \tag{24}$$

At sufficiently low acoustic powers, however, the refractive index profile within the cell is of small amplitude, and the cell will only deflect a small portion of the light. Within this range, the amount of light deflected is roughly linear with acoustic signal. As the acoustic power is increased, more of the light is deflected, and on further increase a saturation point is reached where almost 100% of the light is deflected. No further increase in the intensity of the deflected light is achieved if the acoustic power is increased beyond saturation. By altering the RF amplitude, the intensity of the light is modulated. By altering the RF frequency, the angle of deflection can be varied according to $\Lambda = V/f$, where V is the velocity of the acoustic wave.

If the acoustic signal consists of frequency components within a band given by $f_0 + \Delta f$, the angular divergence of the optical beam is given by

$$\delta\theta = \frac{(2\pi/V)\,\Delta f}{2\pi/\lambda} = \frac{\lambda\,\Delta f}{V} \tag{25}$$

If the optical beam is of angular width $\delta\theta \approx \lambda/D$, where D is the diameter of the beam, then the bandwidth is

$$\Delta f = \frac{V}{D} \tag{26}$$

A higher bandwidth, then, can be achieved with a smaller spot size.

The time-bandwidth product of an acousto-optic modulator is the product of the transit time, τ, across the light beam aperture and the modulator bandwidth:

$$TB = \tau\,\Delta f \tag{27}$$

$$= \frac{D}{V}\,\Delta f \tag{28}$$

The number of resolvable spots, N, on the output of the switch/modulator is defined as the angular spread of the diffracted light divided by the angular spread of the incident light:

$$N = \frac{(\lambda/V)\,\Delta f}{\lambda/D} = TB \tag{29}$$

and thus the number of resolvable spots equals the time-bandwidth product. If the switching time of the RF driver is small, the transit time, τ, of the acoustic wave across the spot diameter can be considered the switching time of the acousto-optic device.

The time-bandwidth product chosen will then depend on the particular application. For a multiple-port switch, a large value of switching ports N can be accommodated by using a large beam aperture. This is accomplished, however, at the expense of a decrease in switching speed.

The highest switch speed would be for a TB product equal to unity, with a small optical aperture only switching into one output port. However, the commercial value of a single-port modulator is limited due to competition from electro-optic modulators, which are covered in Chapter 8. Nevertheless, in contrast to e-o devices, acousto-optic deflectors are capable of switching a beam of light into a large number of ports at moderate speed (less than microseconds) and with low electrical drive power per port. Consequently, the major application of acousto-optic modulators and beam deflectors is in multiport switching.

A difficulty exists in the scannable deflector, in that the exit angle of the light from the device changes with incident angle. Thus, as the deflected beam changes with drive frequency, the incident angle must be altered. This can be done either by rotating the acousto-optic cell with respect to the incident light beam or by rotating the direction of the acoustic wave within the cell. The latter requires an array of transducers along the edge of the cell, which alters the propagation direction of the acoustic wave.

An acousto-optic spectrum analyzer can be fabricated to take advantage of the variation of the light with acoustic frequency. Each component of the acoustic signal will route a disperse incident optical beam into a different output direction. By monitoring the output intensity of each deflected light beam, a spectral analysis of the acoustic signal can be obtained.

An electronically tunable filter can be prepared in certain acousto-optic crystals. In an optical anisotropic medium of these crystals, at a certain orientation, the polarization of the Bragg-diffracted light is orthogonal to that of the incident light. For a certain acoustic frequency only a small range of optical wavelengths will be diffracted into the orthogonal polarization. If the acoustic frequency is varied, the band of optical wavelengths changes, and thus can be scanned.

5 CORRELATION AND CONVOLUTION

Figure 8 illustrates how an acousto-optic cell operating in the Raman-Nath mode can be used as a phase modulator to modulate the wavefront of plane waves. For the moment, we ignore the diffraction angles θ. An electrical signal, usually RF, is sent into a transducer mounted on the acousto-optic device. This sends a strain field into the acousto-optic cell, which is given by

$$s(x - Vt) \tag{30}$$

where V is the velocity of the strain field. Collimated monochromatic light is incident on the device from the left. As the light passes through the cell, it is locally advanced or retarded in phase by the variations in refractive index of the medium. The RF signal is "written," or modulated onto, the wavefront as it exits, and diffracts. If the incident light field is given

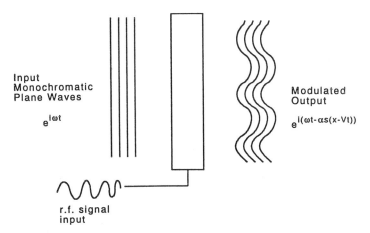

Figure 8 Raman-Nath cell being used as a phase modulator. The RF signal fed into the cell's transducer produces a strain field $s(x - Vt)$ that is written on the wavefront of an incident monochromatic light wave.

by $E_0 e^{i\omega t}$, the electric field of the light emerging is

$$E_0 e^{i[\omega t - \alpha s(x - Vt)]} \tag{31}$$

where α is the modulation index of the RF wave as written onto the optical beam. Usually, α is small and Eq. 31 can be written as

$$E_0 e^{i\omega t}[1 + i\alpha s(x - Vt)] \tag{32}$$

If the RF signal is a simple, modulated cosine wave, as

$$s(x - Vt) = m(x - Vt)\cos[K(x - Vt)] \tag{33}$$

where $m(x - Vt)$ is the amplitude modulation index of the signal as written onto the RF (cosine) wave, the electric field E_{out} of the light exiting the acousto-optic cell is given by

$$E_{out} = E_0 e^{i\omega t}\{1 + i\alpha m(x - Vt)\cos[K(x - Vt)]\} \tag{34}$$

$$= E_0 e^{i\omega t} \left\{ 1 + \frac{i\alpha m(x - Vt)}{2} e^{iK(x - Vt)} + \frac{i\alpha m(x - Vt)}{2} e^{-iK(x - Vt)} \right\} \tag{35}$$

In order to demodulate the carrier, the imaginary terms within the braces of Eq. 35 must somehow be separated and converted to real terms.

Effects of diffraction on the phase-modulated output is shown in Fig. 9. Light is diffracted by the acousto-optic modulator in the $+\theta$ and $-\theta$ directions. If the frequency of the acoustic wave is given by Ω, those waves of the $+\theta$ direction are Doppler-shifted by an amount $\omega + \Omega$, while those diffracted in the $-\theta$ direction travel with a frequency $\omega - \Omega$. These sidebands of the center frequency, ω, are given by

$$E_0 e^{i\omega} \frac{i\alpha m(x - Vt)}{2} e^{iK(x - Vt)} \tag{36}$$

for the $\omega + \Omega$ sideband, and by

$$E_0 e^{i\omega} \frac{i\alpha m(x - Vt)}{2} e^{-iK(x - Vt)} \tag{37}$$

for the $\omega - \Omega$ sideband. If the RF signal is modulated, the modulation is still contained in the $+\theta$ and $-\theta$ diffracted plane waves. If the bandwidth is narrow, the light behaves as an optical carrier with nonoverlapping sidebands, with the sidebands traveling at angles to the optical carrier.

Usable driving frequencies of the cell in Fig. 8 are limited by the cell size and by the duration of the acoustic waves in the cell. The minimum useful time window in the cell is a few tens of microseconds, which limits applications to radio frequencies.

Information contained in the optical carrier is contained in its phase, as according to Eq. 34:

$$E = E_0 e^{i\omega t}\{1 + i\alpha m(x - Vt)\cos[K(x - Vt)]\} \tag{38}$$

which indicates, since the $\alpha m \cos[K(x - Vt)$ term is imaginary, that the

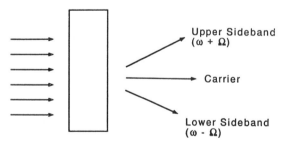

Figure 9 Effects of diffraction on a modulated signal exiting an acousto-optic modulator. Plane waves are separated into an upper sideband of frequency $\omega + \Omega$ traveling at an angle $+\theta$ to a zero-order carrier ω and a lower sideband of frequency $\omega - \Omega$ at an angle of $-\theta$ to the carrier.

modulated component is 90° out of phase with the light wave. Any attempt at "direct detection" of the light using a photodetector followed by a low-pass filter will result in a signal that is proportional to $|E^*E| = 0$, and the modulated information is lost. An attempt must be made, therefore, to convert phase modulation into amplitude or frequency modulation.

One way to convert the phase-modulated signal into a frequency modulated signal is to use a spatial filtering system as shown in Fig. 10. Again, the cell is illuminated with a collimated beam of light. In the back of focal plane of lens L_1 appears the spatial Fourier transform of the distribution of the acoustic stress field $s(x - Vt)$. At this point the carrier and sideband spectra are separated in space, and filtering can be used to select spatial frequencies. A second lens will retransform the light from this plane and an image of the strain field, still phase-modulated, then appears at the image plane of lens 2. Any of three different filtering operations within the transform plane can produce an intensity-modulated, rather than phase-modulated, image at the image plane:

1. Spatial filtering in the image plane, to block one or the other sideband. The resulting sideband can be obtained with a component of amplitude modulation.
2. Delay one-half of the transform plane by 180° by insertion of a half-wave plate in the transform plane.

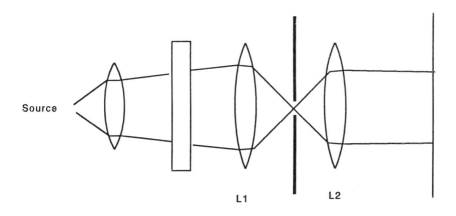

Figure 10 Use of a spatial-filter to convert phase-modulated light carrier exiting an acousto-optic modulator into an intensity-modulated carrier.

3. Delay the carrier relative to the sidebands by insertion of a quarter-wave plate.

For example, consider a phase-modulated signal which has just exited the acousto-optic cell. The signal is passed through a reference modulator whose amplitude transmittance is given by $r_0 + r(x)$. The carrier is then given by

$$E = E_0 e^{+i\omega t}[1 + i\alpha s(x - Vt)][r_0 + r(x)] \tag{39}$$

A spatial filter is inserted in order to block one or the other sidebands in the transform plane. If the $+1$ sideband is blocked, the output intensity of the light is

$$I(t) = \frac{1}{2} \text{Re} \int |E|^2 \, dx \tag{40}$$

where

$$E = E_0 e^{i\omega t} \left\{ 1 + \frac{i\alpha m(x - Vt)}{2} e^{-iK(x - Vt)} \right\} [r_0 + r(x)] \tag{41}$$

and

$$|E|^2 = E_0^2 \left\{ 1 + \frac{\alpha^2 m^2}{4} + \alpha m \sin[K(x - Vt)] \right\} [r_0 + r(x)] \tag{42}$$

If a bandpass filter is inserted after the correlator, low-frequency components will be blocked, and Eq. 40 will yield the correlation of the modulated signal with the reference function:

$$I(t) \propto \int m(x - Vt) r(x) \, dx \tag{43}$$

and the measured intensity is a simple function of the modulation. Similar optical correlators have found applications in real-time radar signal processing [4,5].

6 GUIDED-WAVE ACOUSTO-OPTICS

There are certain advantages in containing the acoustic wave within a piezoelectric waveguide fabricated on a piezoelectric or nonpiezoelectric substrate. This is essentially combining surface acoustic wave (SAW) technology with guided-wave optics to produce a surface wave whose penetration depth reaches the effective width of the channel waveguide.

The waveguide may be a graded-index guide created by diffusing impurities in a substrate such as LiNbO$_3$. Alternatively, a step-index waveguide can be formed by depositing a film on a substrate by a deposition process. In the latter case the substrate does not have to be a good piezoelectric. An example is the deposition of ZnO or As$_2$S$_3$ films formed by sputter deposition on substrates such as silicon. Interdigital electrodes are formed on the surface of the waveguide, and acoustic waves are then introduced into the films as surface acoustic waves. The same optical devices described for bulk acousto-optics can be fabricated in guided-wave a–o devices, including devices for switching, modulation, and optical signal analysis.

One obvious advantage of guided-wave acousto-optics is the possible integration of acousto-optic devices with other optical components, including optical sources, detectors, modulators, etc. One could envision different SAW devices (filters, oscillators, correlators, etc.) fabricated with sources such as laser diodes in an optical integrated circuit. The sources and detectors would be mounted externally to the waveguide. In the case of piezoelectrics such as LiNbO$_3$ that are also electro-optic, both acousto-optic and e–o devices could be included. Waveguide lenses can also be fabricated for the integration of passive and active components on a single substrate.

Other advantages of guided-wave over bulk acousto-optics include smaller size, lighter weight, smaller drive power requirements, and wider bandwidths. Waveguide modulators, for example, require anywhere from one to three orders of magnitude less drive power per unit bandwidth than do their corresponding bulk counterparts.

6.1 Surface Acoustic Wave

A SAW device is shown in Fig. 11. Here, the acoustic wave travels across the surface rather than through the bulk of the device. Interdigital electrodes are placed on the surface of a piezoelectric material, and these act as the source of the acoustic excitation. The acoustic wavelength is determined by the spacing between the interdigital electrodes.

The SAW is the source used for guided-wave acousto-optics. As it is exciting only acoustic waves on the surface of the device, the interaction depth of the acoustic wave will normally not equal the thickness of the optical waveguide. Furthermore, the acoustic wave must interact with an excited mode of the guided optical wave. This is a slightly different situation from the bulk acousto-optic device, in that for the waveguide device neither acoustic nor optical waves have uniform intensity distributions

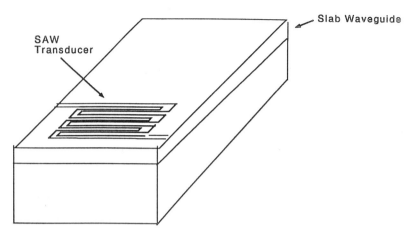

Figure 11   Interdigital electrodes which create surface acoustic waves on a guided-wave acousto-optic device.

within the waveguide. The energy distribution of the acoustic wave will taper off exponentially within the depth of the waveguide (similar to the exponential decrease of the evanescent component of the optical energy distribution). The diffracting power of the device will depend on the overlap factor, ξ, which is defined as [6]

$$\xi = \frac{\text{Transverse energy distribution in the guided optical mode}}{\text{Transverse energy distribution in the surface acoustic wave}}$$

$$\approx \frac{L}{\Lambda} \tag{44}$$

where L is the thickness of the waveguide and Λ is the wavelength of the acoustic wave. If the waveguide is made smaller than the acoustic signal wavelength, ξ will be small and the amount of power necessary to diffract the same amount of light will increase, thus diminishing the advantages of using the waveguide device.

Acousto-optic-mode couplers may also be made, which are similar to the electro-optic devices to be discussed in Chapter 8. The acousto-optic effect is used to affect a localized change in the refractive index of the guide. This refractive index perturbation causes two or more modes within the guide to couple, or a single mode to couple into or out of the waveguide. For example, mode conversion between different transverse electric, or from transverse electric to transverse magnetic, modes is possible.

REFERENCES

1. B. E. A. Saleh and M. C. Teich, *Fundamentals of Photonics*, Wiley, 1991, Chapter 20.
2. G. Wade, Bulk wave acousto-optic Bragg diffraction, in *Guided Wave Acousto-Optics- Interactions, Devices, and Applications* (C. S. Tsai, ed.), Springer-Verlag, 1990, p. 11.
3. N. V. Uchida and N. Niizeki, Acoustooptic deflection, materials and techniques, *Proc. IEEE, 61*, 1073 (1973).
4. W. T. Maloney, Acoustooptical approaches to radar signal processing, *IEEE Spectrum, Oct.*, 40 (1969).
5. W. T. Rhodes, Acousto-optic signal processing: convolution and correlation, *Proc. IEEE, 69*, 65 (1981).
6. J. M. Hammer, Modulation and switching of light in dielectric waveguides, in *Integrated Optics*, 2nd ed. (T. Tamir, ed.), Springer-Verlag, 1979, p. 139.

8

Electro-Optic and Electroabsorptive Guided-Wave Devices

1 INTRODUCTION

There are many advantages for devices integrated in slab waveguides over their use as bulk components. Many of these advantages were pointed out in the integrated acousto-optic devices of Chapter 7. Optical field densities within a small waveguide are much larger for a relatively smaller total optical power than for the bulk counterpart. The devices can be better integrated with optical emitters or switches if waveguides are used to direct the optical signals. For optical devices dependent upon interference effects, diffractionless propagation in a waveguide leads to longer effective interaction lengths than in comparative bulk devices. And finally, by confining the light to a small volume, the number of excitable modes within the device can be controlled.

A wide variety of optical waveguide devices have been proposed and demonstrated, several of which are already on the market. Some of these are purely passive devices, which serve strictly as directional guides or reflectional waveguides. Modulators and switches in integrated optics perform functions analogous to electronic transistors in integrated circuits, and may serve as active components. This means they are able to deliver power to another part of the circuit, rather than merely consuming power. For this chapter we will focus on active modulators and switches based on either the electro-optic or the electroabsorptive effect.

Optical devices can be prepared in optical waveguides in electro-optic single crystals. Different techniques can be used to prepare slab waveguides with defined waveguide patterns. A strip can be delineated in the surface by lithography, as shown in Fig. 1a. More commonly, an impurity may be selectively diffused into a substrate such as titanium into lithium niobate (Fig. 1b). Waveguides can also be prepared by ion implantation of a damaged layer deep into the substrate; the damaged region has a slightly reduced refractive index and thus forms a waveguide with the undamaged region (Fig. 1c).

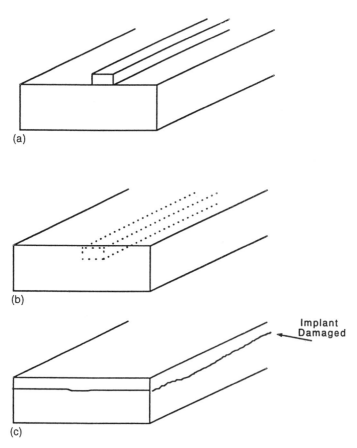

Figure 1 Common methods for preparing optical slab waveguides include (a) etched ridge, (b) diffused planar, and (c) ion implantation.

1.1   The Complex Refractive Index

Electro-optic and electroabsorption devices both operate on the concept of control of the refractive index or change in the refractive index with an applied electric field. It is instructive to review the basics on the complex refractive index.

The propagation, or phase velocity v, of a single wave given by

$$\mathscr{E}(x, t) = \mathscr{E}_0 \exp\{i(kx - \omega t)\} \tag{1}$$

is related to the refractive index by

$$v = \frac{c}{n} = \frac{\omega}{k} \tag{2}$$

To account for the fact that the propagation velocity is frequency dependent, the refractive index is complex:

$$n = n_1 + i n_2 \tag{3}$$

Substituting into Eq. 1 yields

$$\mathscr{E}(x, t) = \mathscr{E}_0 \exp\{-i\omega t\} \exp\left\{\frac{i\omega x n_1}{c}\right\} \exp\left\{\frac{\omega n_2 x}{c}\right\} \tag{4}$$

where the latter term represents a damping factor and a loss in the system power. The imaginary part, n_2, of the refractive index is often referred to as the extinction coefficient and is related to the absorption coefficient by

$$\alpha = \frac{2\omega n_2}{c} \tag{5}$$

α is often treated as the imaginary part of the propagation constant:

$$k = \beta + i\alpha \tag{6}$$

For a material such as a semiconductor of electrical conductivity σ, the propagation of light can be described by the wave equation

$$\frac{\partial^2 \mathscr{E}}{\partial x^2} = 4\pi\sigma \frac{\mu}{c^2} \frac{\partial \mathscr{E}}{\partial t} + \frac{\mu\epsilon}{c^2} \frac{\partial^2 \mathscr{E}}{\partial t^2} \tag{7}$$

where, for convenience, we have written the wave equation in cgs units. When the solution is substituted into the equation, we obtain

$$\frac{\omega^2}{v^2} = \frac{\mu\epsilon\omega^2}{c^2} - \frac{i\omega\mu}{c^2} 4\pi\sigma \tag{8}$$

for semiconductors, $\mu \approx 1$, and we can write

$$\frac{1}{v^2} = \frac{\epsilon}{c^2} - i\frac{4\pi\sigma}{\omega c^2} \tag{9}$$

$$= \frac{n^2}{c^2} \tag{10}$$

If we square the refractive index

$$n^2 = (n_1)^2 - i2n_1n_2 - (n_2)^2 \tag{11}$$

and equate the real and imaginary parts, we obtain expressions for the dielectric constant:

$$(n_1)^2 - (n_2)^2 = \epsilon_1 \tag{12}$$

$$2n_1n_2 = \frac{4\pi\sigma}{\omega} = \epsilon_2 \tag{13}$$

where ϵ_1 and ϵ_2 are the real and imaginary parts of the complex dielectric constant. It is easily shown that

$$(n_1)^2 + (n_2)^2 = \left[\epsilon_1^2 + \left(\frac{4\pi\sigma^2}{\omega^2}\right)^2\right]^{1/2} \tag{14}$$

which gives us, with Eq. 12,

$$(n_1)^2 = \frac{1}{2}\epsilon_1\left\{\left[1 + \left(\frac{\epsilon_2}{\epsilon_1}\right)^2\right]^{1/2} + 1\right\} \tag{15}$$

and

$$(n_2)^2 = \frac{1}{2}\epsilon_1\left\{\left[1 + \left(\frac{\epsilon_2}{\epsilon_1}\right)^2\right]^{1/2} - 1\right\} \tag{16}$$

When the material conductivity approaches zero and/or the frequency of the radiation becomes large, then $\epsilon_2 \to 0$ and $n_1^2 \to \epsilon_1$, $n_2 \to 0$. According to Eq. 5, $\alpha \to 0$ and the material becomes transparent. This explains, at least in a heuristic manner, why good conductors are usually opaque.

Kramers-Kronig Relations
The real and imaginary parts of the complex dielectric function depend upon the frequency of the electromagnetic wave and are related by the Kramers-Kronig relation:

$$\epsilon_1(\omega) = 1 + \frac{2}{\pi}P\int_0^\infty \frac{\omega'\epsilon_2(\omega')\,d\omega'}{[\omega']^2 - \omega^2} \tag{17}$$

$$\epsilon_2(\omega) = -\frac{2\omega}{\pi} P \int_0^\infty \frac{\omega' \epsilon_1(\omega')\, d\omega'}{[\omega']^2 - \omega^2} \tag{18}$$

where P is the principal part of the Cauchy integral:

$$P \int_0^\infty = \lim_{a \to 0} \left(\int_0^{\omega - a} \int_{\omega + a}^\infty \right) \tag{19}$$

Data on the spectral response of the optical absorption can then be used (using Eq. 17) to determine the dispersion in the refractive index, at least over a short range of frequencies.

From Kramers-Kronig, the sum rules for the relative refractive index and dielectric constant can be derived:

$$\int_0^\infty (n_1(\omega) - 1)\, d\omega = 0 \tag{20}$$

which implies that the average value of n_1 over the whole frequency spectrum is unity, and

$$\int_0^\infty (\epsilon_1(\omega) - 1)\, d\omega = -2\pi\sigma_0 \tag{21}$$

implies that for a conductor the average value of the real part of the dielectric constant over the entire frequency spectrum is related to the DC conductivity, σ_0. This is another of the basic relationships between conductivity, n_1, and ϵ_1.

1.2 Model for the Polarization of Solids

If the polarization induced by the optical radiation can be modeled, a better understanding of the complex dielectric function and refractive index can be obtained. Such a model is the damped harmonic oscillator, a model that has been applied to a variety of situations, ranging from nonlinear optics to quantum effects. We start with the one-dimensional equation of motion for a damped harmonic oscillator, which represents the motion of a bound electron under the influence of an oscillating electric field:

$$\frac{d^2x}{dt^2} + \gamma \frac{dx}{dt} + \omega_0^2 x = \frac{q}{m} E_0 \exp\{-i\omega t\} \tag{22}$$

The right-hand side of the equation represents the forced motion of the radiation. On the left side, ω_0 is the natural resonant frequency of the oscillator, and γ is the damping constant. The strength of the damping is determined by how tightly the electron is bound. The solution to this equation of motion has the form

$$x = \frac{(qE_0/m)\exp\{-i\omega t\}}{\omega_0^2 - \omega^2 - i\gamma\omega} \tag{23}$$

The oscillator polarizability, α, can be defined as

$$\alpha = \frac{xq}{E_0} = \frac{q^2/m}{\omega_0^2 - \omega^2 - i\gamma\omega} \tag{24}$$

If there are N electrons per unit volume, the polarizability will be αN. From the Clausius-Mossotti relation, the dielectric constant is related to the polarizability:

$$\epsilon = 1 + \frac{\alpha N}{\alpha N/3 - \epsilon_0}$$

$$= 1 + \frac{N(q^2/m\epsilon_0)}{\omega_0^2 - \omega^2 - i\gamma\omega - Nq^2/3m\epsilon_0} \tag{25}$$

Again, equating the real and imaginary parts yields (from Eqs. 12 and 13)

$$\epsilon_1 = n_1^2 - n_2^2 \tag{26}$$

$$= 1 + N\left(\frac{q^2}{m\epsilon_0}\right)\frac{\omega_1^2 - \omega^2}{(\omega_1^2 - \omega^2)^2 + \gamma^2\omega^2} \tag{27}$$

and

$$\epsilon_2 = 2n_1 n_2 \tag{28}$$

$$= N\left(\frac{q^2}{m\epsilon_0}\right)\frac{\omega\gamma}{(\omega_1^2 - \omega^2)^2 + \gamma^2\omega^2} \tag{29}$$

where a new resonance frequency,

$$\omega_1 = \left[\omega_0^2 - \frac{q^2 N}{3m\epsilon_0}\right]^{1/2} \tag{30}$$

is obtained from the forced damping.

The absorption will be a maximum where n_2 is a maximum, that is, for $\omega = \omega_1$. Differentiating Eq. 29 shows that the conductivity is a maximum at $\omega = \omega_0$. The real part, n_1, increases rapidly with decreasing ω close to ω_0. Eventually, n_1 passes through a maximum, then falls at a rate slower than the increase on the short-wavelength side of the maximum. At each absorption band the value of n_1 then goes through a similar maximum.

This behavior is similar for absorption types other than purely electronic. Figure 2 shows the variation of n_1 and n_2 with excitation energy

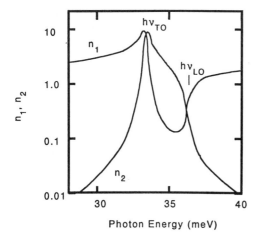

Figure 2   Spectral variation of the real (n_1) and the imaginary (n_2, also known as the extinction coefficient) values of the refractive index as a function of the incident photon energy for the energy range corresponding to polar optical phonon absorption in GaAs. (From Ref. 1.)

$\hbar\omega$, for the range of energies corresponding to the absorption of optical phonons in GaAs. The absorption bands due to the transverse ($h\nu_{TO}$) and longitudinal ($h\nu_{LO}$) polar optical bands can be observed at energies where the extinction coefficient and real refractive index are at a maximum.

For wavelengths far from the absorption maximum, n_2 becomes small compared to n_1 and Eq. 29 can be approximated by

$$n_1^2 - 1 = \frac{N(q^2/m\epsilon_0)}{\omega_0^2 - \omega^2} \tag{31}$$

Thus, by plotting $1/(n_1^2 - 1)$ against ω^2 for ω far from the absorption band, values for ω_0 can be obtained. With this, the low-frequency refractive index can be found.

1.3   The Electro-Optic Effect

In the preceding model, we have assumed that the electron is harmonic in its potential and there are no nonlinear restoring forces. To introduce the electro-optic effect, we now rewrite Eq. 22 to include the effects of a nonlinear potential:

$$\frac{d^2x}{dt^2} + \gamma\frac{dx}{dt} + \omega_0^2 x + g_2 x^2 + g_3 x^3 = \frac{q}{m} E_0 \exp\{-i\omega t\} \tag{32}$$

where the term g_2 is responsible for the second-order nonlinearities, including the electro-optic (Pockels) effect and second-harmonic generation, and the term g_3 is the constant which governs the third-order nonlinear effects, such as the optical Kerr effect.

Normally, the g_3 term is weak, and the second-order nonlinear effect governed by g_2 will dominate. The g_2 term is only pertinent for noncentrosymmetric crystals, that is, crystals lacking a center of inversion symmetry. For a sinusoidal driving electric field represented by the right-hand side of Eq. 32, the second-order effect will produce harmonics around the central frequency ω_0, with frequencies spaced at integral values of ω. For nonlinear driving fields, the anharmonic terms of the polarization will mix with the nonlinear terms of the driving field to produce harmonics at sum and difference frequencies.

If we ignore the terms in g_3, a solution for Eq. 32 can be written as

$$x(t) = x_0 + \frac{1}{2}(x_1 e^{i\omega t}) \tag{33}$$

Substituting Eq. 33 into Eq. 32 gives solutions for the constants x_1 and x_2. If an applied electric DC field, E_{DC} is added, the solution can be separated into a DC term

$$\frac{\omega_0^2}{2} x_0 + g_2 \left[x_0^2 + \frac{x_1^2}{2} \right] = \frac{q}{m} E_{DC} \tag{34}$$

and an AC term

$$x_1 [\omega_0^2 - \omega^2 - i\gamma\omega + 2g_2 x_0] = \frac{q}{m} E_0 \tag{35}$$

and the polarizability is now

$$\alpha = \frac{q^2/m}{\omega_0^2 - \omega^2 - i\gamma\omega + 2g_2 x_0} \tag{36}$$

and the $g_2 x_0$ term now has a direct effect on the polarizability.

The effect of the applied field is to shift the origin of the oscillation from $x = 0$ to $x = x_0$. The value of this origin is dependent upon the value of E_{DC}. Because of the nonlinearity, the frequency of oscillation about $x = x_0$ is different from that about $x = 0$, altering the polarizability and, hence, the refractive index. The refractive index becomes dependent upon the applied field through the dependency of the $2g_2 x$ term of the polarizability.

For the electro-optic effect to be effective in an electro-optic device, the materials of device construction must lack the all-important center of inversion symmetry. This lack of symmetry causes the dielectric properties to vary as a function of direction within the crystal. Anisotropic dielectric properties can be analyzed from the directional-dependent dielectric constant, ϵ_{ij}, which determines the relation between the electric flux density D_i and the electric field strength E_i:

$$D_i = \epsilon_{ij}\epsilon_0 E_j \tag{37}$$

where ϵ_{ij} are the *relative* dielectric constants. If the dielectric inhomogeneity of the crystal is aligned with the orthogonal coordinate system, ϵ_{ij} can be written as a third-ranked tensor with off-diagonal components equal to zero:

$$|\epsilon| = \begin{vmatrix} \epsilon_{11} & & \\ & \epsilon_{22} & \\ & & \epsilon_{33} \end{vmatrix} \tag{38}$$

The stored energy in the crystal, w, is

$$w = \frac{1}{2}\mathbf{E}\cdot\mathbf{D} \tag{39}$$

Substitution of Eqs. 37 and 38 into 39 yields

$$w = \frac{1}{2\epsilon_0}\left(\frac{D_x^2}{\epsilon_{11}} + \frac{D_y^2}{\epsilon_{22}} + \frac{D_z^2}{\epsilon_{33}}\right) \tag{40}$$

Setting

$$x = \frac{D_x}{(2\epsilon_0 w)^{1/2}}, \qquad y = \frac{D_y}{(2\epsilon_0 w)^{1/2}}, \qquad z = \frac{D_z}{(2\epsilon_0 w)^{1/2}} \tag{41}$$

and

$$\epsilon_{ij} = n_i^2 \tag{42}$$

gives an equation for an ellipsoid:

$$\frac{x^2}{n_x^2} + \frac{y^2}{n_y^2} + \frac{z^2}{n_z^2} = 1 \tag{43}$$

The index ellipsoid, or indicatrix, describes the variation of the refractive index with direction in the crystal. The directions x, y, and z are the *principal axes* or *privileged directions* of the crystal.

The privileged directions are associated with *birefringence*, which is the naturally occurring anisotropic variation of the refractive index in noncentrosymmetric compounds. For centrosymmetric cubic crystals under no applied electric field, or any other external disturbance, $n_x = n_y = n_z$, the indicatrix is a sphere, and the material is nonbirefringent. For materials of tetragonal, trigonal, or hexagonal symmetry, the index ellipsoid is symmetric about the z-axis, with $n_x = n_y \neq n_z$.

Perturbations such as an applied electric field can either alter the natural birefringence of a crystal or enhance birefringence in a non- or weakly birefringent solid. Such solids are considered *electro-optic*. The perturbation acts to distort the indicatrix and affect the birefringence.

Figure 3 is a representation of an indicatrix in an electro-optic solid. The z and x (or z and y) directions in the figure are the privileged directions, and the z is also the optical axis, as shown. In this representation, $n_x = n_y \neq n_z$, and the values of n_e and n_0 define the indicatrix.

Light incident at an oblique angle to the crystal is represented in Fig. 3 as propagating along the direction OB. The electric field vector of the propagating radiation can be decomposed into two components, E_0 and E_e. E_0 is perpendicular to the z-axis, while E_e is perpendicular to E_0. The E_0 component will travel with a velocity c/n_0, while the E_e travels at

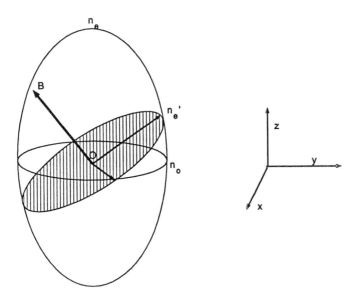

Figure 3 Index indicatrix of a uniaxial birefringent crystal. The three-dimensional shape of the indicatrix represents the variation of refractive index with direction in the crystal. For light propagating along the direction OB, the electric field vector is decomposed into two components which will have different velocities within the crystal.

c/n'_e, where n'_e is determined by the component of the indicatrix along the E_e direction. The possible values of n'_e are represented by the shaded region of Fig. 3. The two components of the electric field of the incident radiation, traveling at different velocities, will separate. Two beams of light will then be traveling within the crystal at two different velocities. These crystals are sometimes called doubly diffracting; since they have different values of refractive index they refract at different refractive angles. Light propagating parallel to the z, or optical, axis will not separate into individual components since, along that direction, $n'_e = n_0$.

Switching and modulating devices can be fabricated as optoelectronic elements in bulk electro-optic solids or as slab optical waveguides formed in electro-optic crystals. Such waveguides can be fabricated in doubly refracting crystals such as KDP (potassium dihydrogen phosphate, KH_2PO_4) or lithium niobate ($LiNbO_3$). These crystals are double refracting since the index of refraction, and thus the speed of propagation, is a function of at least two different directions in the crystal. Furthermore, the differences in the index for the two directions in both naturally isotropic and naturally double refracting (or birifringent) can be enhanced via the linear electro-optic effect.

In the presence of an applied electric field, the index ellipsoid can be expressed as

$$B_{11}x^2 + B_{22}y^2 + B_{33}z^2 + 2B_{12}xy + 2B_{23}yz + 2B_{31}zx = 1$$

The axes x, y, and z are no longer the privileged directions, and the coefficients B_{ij} are dependent upon the applied electric field according to

$$\begin{bmatrix} B_{11} - 1/n_x^2 \\ B_{22} - 1/n_y^2 \\ B_{33} - 1/n_z^2 \\ B_{23} \\ B_{31} \\ B_{12} \end{bmatrix} = \begin{bmatrix} r_{11} & r_{12} & r_{13} \\ r_{21} & r_{22} & r_{23} \\ r_{31} & r_{32} & r_{33} \\ r_{41} & r_{42} & r_{43} \\ r_{51} & r_{52} & r_{53} \\ r_{61} & r_{62} & r_{63} \end{bmatrix} \begin{bmatrix} E_x^{DC} \\ E_y^{DC} \\ E_z^{DC} \end{bmatrix} \tag{44}$$

where E_x^{DC}, E_y^{DC}, E_z^{DC} are the vector components of the applied field, and r_{ij} are the electro-optic coefficients. In the case of a crystal such as $LiNbO_3$, symmetry reduces the electro-optic tensor to

$$\begin{bmatrix} 0 & -r_{22} & r_{13} \\ 0 & r_{22} & r_{13} \\ 0 & 0 & r_{33} \\ 0 & r_{51} & 0 \\ r_{51} & 0 & 0 \\ -r_{22} & 0 & 0 \end{bmatrix} \tag{45}$$

Measured values of the coefficients r_{13}, r_{22}, r_{33}, and r_{51} for LiNbO$_3$ are 9.6×10^{-12}, 6.8×10^{-12}, 30.9×10^{-12}, and 32.6×10^{-12} m/V, respectively. The crystal is uniaxial, with $n_x = n_y = n_0$, and $n_z = n_e$.

For III-V compound semiconductors which are cubic, but lack a center of inversion symmetry, the tensor is

$$\begin{bmatrix} 0 & 0 & 0 \\ 0 & 0 & 0 \\ 0 & 0 & 0 \\ r_{41} & 0 & 0 \\ 0 & r_{41} & 0 \\ 0 & 0 & r_{41} \end{bmatrix} \tag{46}$$

The value of r_{41} is not large, on the order of 10^{-12} m/V, for most III-V compounds. However, the large refractive index (3.6) still makes the electro-optic effect important in semiconductors such as GaAs.

It is strongly desirable to construct electro-optic-based waveguide devices on compound semiconductors, as this would allow complete monolithic optoelectronic integration of sources, detectors, and switches connected by waveguides on a single substrate. Such optical "integrated circuits" would demonstrate remarkable information processing capabilities at high bandwidths with high noise immunity. Unfortunately, the photoelastic properties of GaAs are dependent upon the wavelength of the exciting optical radiation. As the energy of the radiation approaches the band-gap energy of the semiconductor, the value of the electro-optic coefficient increases. However, at the same time, the absorption also increases. This is a fundamental relationship which is governed by the Kramers-Kronig relation. Therefore, there is a balance between high electro-optic effect and low absorption.

For LiNbO$_3$ the constant B_{33} can be found as

$$B_{33} = \frac{1}{n_e^2} + r_{33}E_z^{\mathrm{DC}} \tag{47}$$

or, for a change produced by an electric field along the x axis,

$$\Delta B_{33} = -\Delta \left(\frac{1}{n_e}\right)^2 = r_{33}E_z^{\mathrm{DC}} \tag{48}$$

The effect of applying the electric field is modeled, then, as a perturbation of the impermitivity tensor represented by the diagonal elements $1/n_x^2$, $1/n_y^2$, and $1/n_z^2$. The amount of the perturbation is equal to the value of the electro-optic coefficients.

In the electro-optic effect, an electric field applied to a crystal induces a change in the refractive index. The change in index as a function of this field can then be described by equations similar to Eq. 48:

$$\Delta \left(\frac{1}{n^2}\right) = r\mathcal{E} + P\mathcal{E}^2 \tag{49}$$

where r is the linear electro-optic coefficient and P is the quadratic coefficient. In solids, the linear electro-optic describes *Pockels effect*, while the quadratic electro-optic effect is called the *Kerr effect*. In Pockels effect the applied electric field introduces privileged directions in the crystal which are perpendicular to the applied field. For light entering the crystal, the two components of the polarization which lie along the directions of these privileged directions will then have different optical path lengths.

As an illustration, consider Fig. 4. Incoming plane polarized light, such as produced by laser excitation, is incident on an electro-optic crystal. The light is propagating in the z-direction, and the privileged directions are induced in the x- and y-directions. The light is incident in such a way as the plane of polarization perfectly bisects the x- and y-axes (the

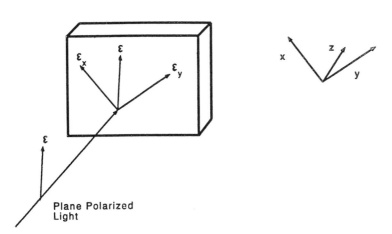

Figure 4 Basic principles of Pockels effect modulator. Plane polarized light is incident on the crystal in an orientation such that the plane of polarization bisects the two privileged directions of the crystal. Since these privileged directions are induced by the applied electric field, the result is control of the x and y components of the electric field vector of the polarized light.

privileged directions). Assuming only Pockels effect is present, an applied electric field in the z-direction will induce a change in the refractive given by Eq. 49:

$$\Delta\left(\frac{1}{n^2}\right) = \frac{-2\Delta n}{n^3} = r\mathscr{E} \tag{50}$$

where Δn represents the difference in refractive index between the two different privileged directions. If this change is small one can write

$$\Delta n = \pm\frac{1}{2}rn_0^3\mathscr{E} \tag{51}$$

where n_0 is the index with no applied field. The x and y coordinates of the refractive index n_x and n_y are determined by

$$n_x = n_0 + \frac{1}{2}rn_0^3\mathscr{E} \tag{52}$$

and

$$n_y = n_0 - \frac{1}{2}rn_0^3\mathscr{E} \tag{53}$$

which corresponds to a phase difference in each direction:

$$\phi_x = \frac{2\pi}{\lambda}n_xL \tag{54}$$

$$\phi_y = \frac{2\pi}{\lambda}n_yL \tag{55}$$

where L is the length of the crystal, and thus a net phase difference (or retardation) of

$$\Phi = \phi_x - \phi_y = \frac{2\pi}{\lambda}rn_0^3V \tag{56}$$

where $V = \mathscr{E}L$ is the applied voltage. If the light emerging from the crystal is such that $\Phi = \frac{\pi}{2}$, then that light is circular polarized. If $\Phi = \pi$, the emerging light is plane polarized, with the plane of polarization at 90° with respect to the polarization plane of the incident light. For any other values of Φ the resulting output of the crystal is elliptically polarized light.

In a commercial Pockels cell modulator, linear polarizing devices are placed on the front and back of the crystal with the plane of polarization 90° to one another. The front polarizer produces the plane polarized light necessary as input for phase retardation by the crystal. The device modulates light by rotating the plane of polarization into the output polarizer when the applied voltage is such that $\Phi = \pi$.

Polarization Preservation

The above-described effects require altering the polarization of the light via the electro-optic effect. Care must then be taken that the polarization is not altered by any other means. A major problem with electro-optic devices is the uniformity in which all components of the polarization may be switched equally. This problem stems from certain electro-optic materials where the electro-optic coefficient is highly anisotropic. In a ferroelectric such as $LiNbO_3$, which has low symmetry, anisotropy in the electro-optic coefficients means that the different polarizations of the optical mode cannot be switched equally. Possible solutions to this problem include preserving the polarization through preserving waveguides, assessing the state of polarization upon switching and correcting for the asymmetry, designing polarization-independent devices, and separating the polarizations and treating each separately. Each method has its place in applied electro-optic devices.

2 ELECTRO-OPTIC EFFECT WAVEGUIDE DEVICES

The electro-optic effect allows a powerful means for the control of an optical beam in a waveguide. These devices can be divided into three basic types, which will be reviewed here:

1. Phase control of guided light can be performed in waveguides similar to how light is controlled in bulk Pockels modulator described above.
2. The interaction length of directional couplers can be modified by an applied voltage to two closely spaced waveguides fabricated in electro-optic materials. Coupling between the waveguides is altered as modeled by coupled-mode theory.
3. The refractive index of the guide, or the surrounding cladding, can be modified. Index control devices can be made to alter the modal structure of one or more waveguides.

2.1 Single-Waveguide Modulators

Figure 5 shows a single-waveguide modulator capable of being used either as a phase or as an amplitude (intensity) modulator. The device consists of a metal gate prepared on top of the structure to form a Schottky diode with the waveguide. A reverse bias is applied to the gate, and the depletion region of the diode becomes part of the waveguide, extending into the

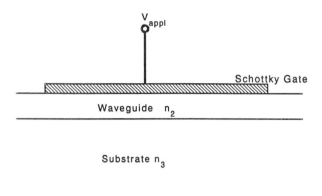

Figure 5 Single-waveguide modulator constructed from a simple Schottky diode.

guide further with increase of the voltage. The waveguide refractive index is modulated by an amount Δn by one or both of two methods:

1. The presence of the depletion region depletes the waveguide of free carriers and the refractive index changes by an amount $\Delta n_{n,p}$, due to the dependence of index on free-carrier density.
2. The presence of the electric field changes the index by an amount Δn_{eo} due to the electro-optic effect.

An intensity modulator can be constructed by designing a waveguide which will just barely support a single mode. When the guide is influenced by the refractive change $\Delta n_{n,p} + \Delta n_{eo}$, the refractive index is altered beyond the threshold for supporting the single mode, and the guide becomes transmissive with the substrate (cladding).

The polarization of free carriers can be studied if the vibrating electrons (or holes) as excited by a sinusoidal electric field is again treated as a damped harmonic oscillator of Section 1.2:

$$m^*_{n,p} \frac{d^2x}{dt^2} + m^*_{n,p}\gamma \frac{dx}{dt} + Kx = q\mathcal{E}_0 \exp\{-i\omega t\} \tag{57}$$

where $m^*_{n,p}$ is now the effective mass of the electron (hole). The wave represented by $\mathcal{E} = \mathcal{E}_0 \exp\{-i\omega t\}$ of Eq. 57 is the exciting electric field. Since free electrons are only weakly bound to the crystal, there is no restoring force in the equation of motion, and the spring constant K is zero. The solution of Eq. 57 for free carriers is then

$$x = \frac{-q\mathcal{E}_0 \exp\{-i\omega t\}}{\omega^2 m_{n,p}^*} \tag{58}$$

and the polarization, α is given by

$$\alpha = \frac{qx}{\mathcal{E}} = \frac{-q^2}{\omega^2 m_{n,p}^*} \tag{59}$$

Because the restoring force is so small, the electrons effectively have no resonant frequency. This makes the denominator of Eq. 59 very large, and it becomes even larger at higher frequencies. This small polarization makes modulation control by free-carriers only feasible at low excitation frequencies.

Since the quantity $\Delta n_{n,p}$ is small, it is usually necessary to use a wave-guide material that possesses a large value for Δn_{eo} to obtain a significant index change. This precludes using materials with small electro-optic coefficients, such as compound semiconductors, for single-waveguide devices.

2.2 Coupled-Mode Devices

Coupled-Mode Theory

A more robust electro-optic device in guided-wave optics relies on the coupling of modes between two or more closely spaced waveguides for operation. For this reason, it is informative to review mode coupling theory.

The formalism of coupled-mode theory has been applied in many instances in science. An electric polarization, $\mathbf{P}$, can be imparted into the waveguide by a deformation in the dielectric of the guide, $\Delta\epsilon$. This deformation may be caused, for example, by a small inhomogeneity in the guide. Such an induced polarization under an electric field $\mathcal{E}$ is given by

$$\mathbf{P} = \Delta\epsilon\mathcal{E} \tag{60}$$

The deformation in the (real) part of the dielectric constant also is represented by a deformation of the (again real) part of the refractive index, Δn. An imaginary value of $\Delta\epsilon$ results in a net loss in the waveguide. Nonlinear optical effects are modeled by a $\Delta\epsilon$ of the form

$$\Delta\epsilon_{ij} = \epsilon_0 X_{ijk}\mathcal{E}_k \tag{61}$$

where the dielectric perturbation is represented as a tensor, as the anisotropy in the guide is usually directional. X_{ijk} is then the second-order nonlinear susceptibility.

Yariv [2] has demonstrated that the electric field in the perturbed waveguide can be expanded as

$$\mathscr{E} = \sum_m A_m(z)\mathscr{E}_m(y)\exp(i\beta_m z) \tag{62}$$

where the waveguide is extending into the direction z, and the direction y is perpendicular to the waveguide. $A_m(z)$ is the complex amplitude of the electric field of mode m. These different modes represented by m are each solutions of the modal equations for the guide as a whole.

Figure 6 illustrates how the perturbation scheme of mode coupling may be modeled. Two modes are considered, whose unperturbed electric field amplitude profiles are given by $\mathscr{E}_1(y)$ and $\mathscr{E}_2(y)$, in a perturbed slab wave-

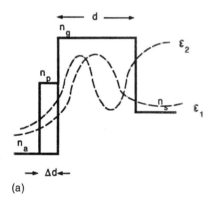

(a)

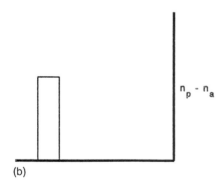

(b)

Figure 6 (a) Modeling the coupling of two modes, whose unperturbed electric field amplitude profiles are given by $\mathscr{E}_1(y)$ and $\mathscr{E}_2(y)$, in a perturbed slab waveguide. The waveguide is of thickness d with a refractive index of n_g in a substrate of index n_s. The perturbation is of thickness Δd and has refractive index n_p. (b) Index perturbation Δn.

guide. The waveguide is of thickness d, with a refractive index of n_g, in a substrate of index n_s. The perturbation is a thin region of the waveguide of thickness Δd and has refractive index n_p. The idea of the model is to solve the electric field for the perturbed case, as a linear solution of the unperturbed $\mathcal{E}_1(y)$ and $\mathcal{E}_2(y)$. Buckman [3] has provided an excellent review on the model, which will be briefly summarized here.

Figure 6b shows the index perturbation, Δn, for the guide in the figure. The index variation has a spatial dependence $\Delta n(y)$ which can be conveniently modeled in terms of the square of the distribution:

$$\Delta n^2(y) = n_0^2(y) - n^2(y) \tag{63}$$

where $n^2(y)$ is the perturbation index distribution on the unperturbed distribution, $n_0^2(y)$. These refractive index distributions give rise to perturbed and unperturbed polarizations. The unperturbed part is written as P_0, and $\Delta n^2(y)$ affects a perturbation component of the polarization P_p, the two of which sum together to give the total polarization P:

$$P = P_p + P_0 \tag{64}$$

Since

$$D = \epsilon_0 \mathcal{E} + P = \epsilon_0 n^2(y)\mathcal{E} \tag{65}$$

then

$$P = \epsilon_0[(n_0^2(y) - 1) + \Delta n^2(y)]\mathcal{E} \tag{66}$$

Examination of Eq. 66 shows clearly the perturbation component of the polarization as

$$P_p = \epsilon_0 \Delta n^2(y)\mathcal{E} \tag{67}$$

$$= \epsilon_0 \Delta n^2(y) \sum_m A_m(z)\mathcal{E}_m(y)\exp(i\beta_m z) \tag{68}$$

The next step is to solve the wave equation by separating the perturbed and unperturbed components.

$$\nabla^2\mathcal{E} + \omega^2\mu_0\epsilon_0 n_0^2(y)\mathcal{E}(y) = \mu_0 \frac{\partial^2 P_p}{\partial t^2} \tag{69}$$

where the perturbed components are separated to the right-hand side of the equation. Using Eq. 62 and assuming a slow change in the field amplitudes $A_m(z)$ with z yields a form

$$\sum_m 2i\beta_m\mathcal{E}_m \frac{\partial A_m(z)}{\partial z} e^{i\beta_m z} = \mu_0 \frac{\partial^2 P_p}{\partial t^2} \tag{70}$$

Using the polarization as given by Eq. 68, and the orthonormality of the field amplitudes, $\mathscr{E}_m(z)$, yields a final result:

$$\frac{dA_n}{dz} - C_n A_n = i \sum_{m \neq n} A_m \kappa_{mn} \exp\{i\Delta_{mn}z\} \tag{71}$$

where

$$C_n = \frac{-k_0^2}{2i\beta_n} \int \mathscr{E}_n^* \Delta n^2 \mathscr{E}_n \, dy \tag{72}$$

$$\kappa_{mn} = \frac{-k_0^2}{2i\beta_n} \int \mathscr{E}_m^* \Delta n^2 \mathscr{E}_n \, dy \tag{73}$$

and

$$\Delta_{mn} = \beta_m - \beta_n \tag{74}$$

The parameter κ_{mn} is a coupling coefficient which represents the strength of the coupling between modes m and n. If κ_{mn} were zero, the power carried in these modes would be constant with z, the direction of propagation. C_n is a measure of how the perturbation has affected the propagation constant β, and Δ_{mn} is a measure of the mismatch of the propagation constants between m and n.

A solution for Eq. 71 can be given as

$$\mathscr{E}_n(z) = A_n(z)\exp\{iC_n z\} \tag{75}$$

where $A_n(z)$ is the z-dependent amplitude of the modified mode n. Substituting this solution back into Eq. 71 yields a set of coupled-mode equations for $A_n(z)$:

$$\frac{dA_n}{dz} = \sum_{m \neq n} A_m \kappa_{mn} \exp\{i\Delta'_{mn}z\} \tag{76}$$

where Δ'_{mn} is given by

$$\Delta'_{mn} = [\beta_m + C_m] - [\beta_n + C_n] \tag{77}$$

As an example, consider two modes, 1 and 2, with complex amplitudes A_1 and A_2. In the unperturbed case, they propagate as single modes, which can be described by

$$\mathscr{E}_1(x, z, t) = A_1 e^{i(\omega_1 t - \beta_1 z)} \tag{78}$$

and

$$\mathscr{E}_2(x, z, t) = A_2 e^{i(\omega_2 t - \beta_2 z)} \tag{79}$$

which obey the relation

$$\frac{dA_1}{dz} = \kappa_{12}A_2 e^{-i\Delta\beta z} \tag{80}$$

and

$$\frac{dA_2}{dz} = \kappa_{21}A_1 e^{-i\Delta\beta z} \tag{81}$$

where $\Delta\beta = \Delta_{12}$. The amplitudes are no longer constant, but are both functions of the propagation direction z.

Directional Couplers

Coupled-mode theory can easily be applied to a directional coupler. Figure 7 illustrates the device. Two waveguides are brought together for a distance L to facilitate mode coupling between the guides along that distance. In this case, for two modes 1 and 2 along each waveguide, we define the two modes in such a way that the power carried by each mode is equal to $|A_1|^2$ and $|A_2|^2$, respectively. Thus, for conservation of total power

$$\frac{d}{dz}(|A_1|^2 + |A_2|^2) = 0 \tag{82}$$

and then

$$\kappa_{12} = -\kappa_{21}^* \tag{83}$$

If the boundary conditions are set such that $A_1(0) = A_{1,0}$ and $A_2(0) = 0$, then solutions are written as

$$A_1(z) = A_{1,0}\frac{\kappa_{12}}{g}e^{-i\Delta\beta z/2}\sin[gz] \tag{84}$$

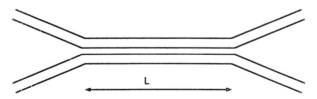

Figure 7   Schematic of a directional coupler, where two waveguides are brought in close proximity for a length L. The waveguides are close enough that evanescent coupling is possible between the waveguides.

and

$$A_2(z) = A_{1,0}e^{-i\Delta\beta z/2}\left\{\cos[gz] - i\frac{\Delta\beta}{2g}\sin[gz]\right\} \tag{85}$$

where

$$g = \frac{1}{2}(4\kappa^2 + \Delta\beta^2)^{1/2} \tag{86}$$

and $\kappa^2 = |\kappa_{12}|^2$.

For Eq. 84 and Eq. 85, maximum coupling between modes occurs for $\Delta\beta = 0$. At this point the solutions become

$$A_1(z, t) = A_{1,0}e^{i(\omega_1 t - \Delta\beta_1 z)} \sin(\kappa z)$$

$$A_2(z, t) = A_{1,0}e^{i(\omega_2 t - \Delta\beta_2 z)} \cos(\kappa z) \tag{87}$$

and the power coupled between waveguides is

$$I_1(z) = |A_1|^2 = I \sin^2(\kappa z) \tag{88}$$

and

$$I_2(z) = |A_2|^2 = I \cos^2(\kappa z)$$

Equation 88 shows that energy then transfers back and forth from one waveguide to the other over the length of the interaction distance L. A $\sin^2$ variation of intensity with distance exists along the length of the waveguide. For a coupler to transfer a fraction of the energy from one waveguide to another, it is only necessary to fabricate the coupler with a proper value of L. No further coupling exists once the light leaves the interaction distance.

There is also a difference in the phase between the modes in the two waveguides. The phase in the driven guide lags that in the driving guide by $\pi/2$. For an interaction distance L_c such that $\kappa L_c = \pi/2$, all of the optical power has been transferred to guide 1. Because of this phase relationship, no energy can be transferred into a backward wave traveling in the $-z$-direction, and hence the coupler behaves as a directional coupler. The interaction length L necessary for complete power transfer from one guide to the other is given by

$$L = L_c(2m + 1), \qquad m = 0, 1, 2, 3, \ldots \tag{89}$$

where

$$L_c = \frac{\pi}{2\kappa} \tag{90}$$

The distance $\pi/2\kappa$ is called the coupling length. For a 3-dB coupler, a field difference of $\kappa L = \pi/4$ would divide such that after the light exits the coupler, the power in the driven wave would be half the power in the driving wave. Complete extinction of the optical power is obtained when

$$L = \frac{\pi}{\kappa}(m + 1), \qquad \text{where } m = 0, 1, 2, \ldots \qquad (91)$$

Electro-optic Directional Coupler/Modulator

To achieve a complete crossover state where 100% of the light crosses from the first guide into the second requires a phase difference $\Delta\beta = 0$ plus an interaction length that is an odd multiple of the coupling length, as per Eq. 89. If the coupler is fabricated in an electro-optic substrate, the phase difference between the guides can be altered through the electro-optic effect. Figure 8 shows the topographical layout of an electro-optic modulator based on a two-channel directional coupler. The device is fabricated with an interaction length L that will provide for maximum coupling with the voltage off (i.e., the device is "normally on"). An applied electric field will shift the relative phase between the guides and the device turns off. For this case, 100% extinction is obtained when $g = \pi/L$. The value of $\Delta\beta$ required for 100% extinction is then given by

$$\Delta\beta = \frac{\sqrt{3}\pi}{L} \qquad (92)$$

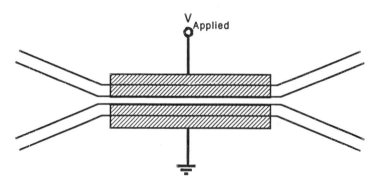

Figure 8 A two-channel electro-optic modulator/switch fabricated from a directional coupler. The coupler is fabricated with an interaction length L that provides for 100% coupling with no applied field. An applied voltage then perturbes the effective index of the guide, and hence the phase in each arm, until light will no longer couple into the second guide.

And the index change needed for 100% extinction is given by

$$\Delta n = \frac{\sqrt{3}\pi}{kL} \tag{93}$$

where k is the wavevector.

Reversed $\Delta\beta$ Modulator

A problem with the straight two-channel modulator of Fig. 8 is that in order to achieve high switching extinction ratios the fabrication tolerance of the device must be very close. There is no way to adjust the on state of the device except by varying the interaction length of the coupler. A two-channel modulator can be fabricated in such a way that the on and off states can be "trimmed" by adjustment of the electrode length rather than the waveguide interaction length. Figure 9 shows a stepped, reversed $\Delta\beta$ coupler, where the electrodes are split in half. Each section of the switch has equal magnitude but reversed phase when a voltage is applied. The on and off states can then be adjusted by adjusting the applied voltage to each half of the switch or by physically trimming the size of the electrodes to achieve a maximum on/off ratio.

2.3   Mach-Zehnder Modulator

A Mach-Zehnder modulator can be fabricated in an electro-optic waveguide in a fashion shown in Fig. 10. Here a XOR/XNOR logic operation can be achieved in the optical domain by applying logic decisions in the electronic domain. Light comes through the waveguide on the left and is split equally in two directions through two separate branches of the

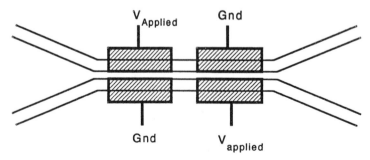

Figure 9   A two-channel directional coupler fabricated using a stepped, reversed $\Delta\beta$ configuration. The configuration allows for easier adjustment of the on state of the coupler as compared to the device in Fig. 8.

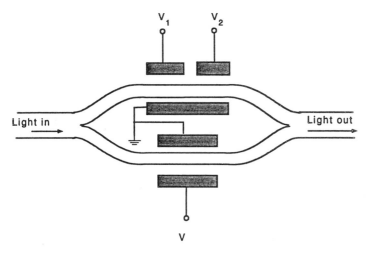

Figure 10 Mach-Zehnder-style logic gate which will execute an XOR/XNOR logic operation.

waveguide. On each side of the waveguide are electrodes which form the switches for the device. A voltage V applied to one switch will introduce a phase shift to that arm. The guided modes are phase-shifted by an amount $+\Delta\phi$ and $-\Delta\phi$ in each branch, which gives a net phase shift of $2\Delta\phi$:

$$2\Delta\phi = \pi \frac{V}{V_{1/2}} \tag{94}$$

where $V_{1/2}$ is the half-wave voltage, or the voltage required to give a phase shift of $180°$ between the waves in each branch. This device is entirely analogous to the bulk phase modulators discussed in Section 1.3. The half-wave voltage from that section is where Eq. 56 is set equal to $\pi/2$:

$$V_{1/2} = \frac{\lambda}{2rn_0^3} \tag{95}$$

If the net phase shift between the two arms is an odd multiple of π, the two light beams cancel where the two arms rejoin, and the output of the device is near zero. The amount of extinction depends on the ideality of the modulator. If the complex amplitudes of the two branches are given by E_1 and E_2, respectively, then the output power where the two are summed together at the output Y branch is given by

$$P_0 = \frac{1}{2} (|E_1| - |E_2|)^2 + 2|E_1||E_2| \cos^2(\Delta\phi) \qquad (96)$$

By changing the voltage it is then possible to control the output power on the device, which varies as a $\cos^2$ function. Figure 11 shows variation of output power with applied voltage. The extinction ratio (output power versus input in the off state) is dependent on how well the light can be equally divided into each branch and on how well the polarization can be maintained within the device. Optoelectronic waveguide devices based on the interferometric design have been constructed which can respond on the order of hundreds of picoseconds.

The above device has a problem of being nonlinear, but the graph represented by Fig. 11 can be shifted by a half-wavelength to give a quasi-linear device. This is accomplished by introducing an initial $\pi/2$ phase shift into one of the branches. Such a phase shift can be made with an asymmetric modulator, where one branch is longer than the other by a quarter-guide wavelength [4].

Instead of a Y-branch, the initial input optical power can be split with a 3-dB directional coupler. Figure 12 illustrates a Mach-Zehnder-type interferometric switch utilizing two such couplers: one to initially split the optical power and one to recombine the two branches. The phase in one branch is again modified by using the electro-optic effect through the electrodes placed on that branch. The light in the two branches is of approximately the same amplitude; only a 90° phase shift exists between the two, as dictated by coupled-mode theory. The electro-optic modulator again acts to alter the phase of one branch against the other before the second coupler recombines the two beams.

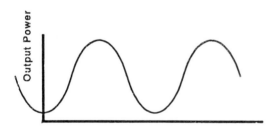

Applied Voltage

Figure 11   Output power of the Mach-Zehnder-type interferometric device vs. applied voltage. The extinction ratio of the switch is dependent on the ability to equally split the optical power between each branch of the device and on the ability to create destructive interference through phase control.

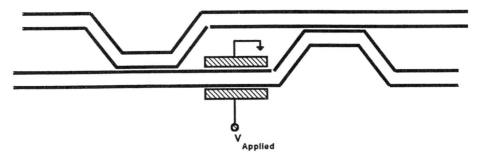

Figure 12 A Mach-Zehnder interferometric modulator fabricated using 3dB couplers to split/recombine the optical power rather than the Y coupler of Fig. 11.

The devices mentioned require precise fabrication tolerances in order to keep the optical path length the same between the two branches. A simpler device is shown in Fig. 13 [5]. This is simply two waveguides that are fabricated in such close proximity that the beam in each guide will undergo evanescent coupling. However, a narrow groove is machined between the waveguides to forbid coupling in that region. Enough coupler

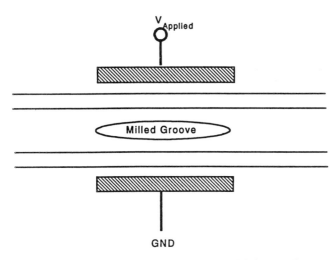

Figure 13 An interferometric switch fabricated from a single evanescent coupler.

is left on each end of the groove to allow for two 3-dB couplers needed for the interferometer. The electrodes are placed on each side of the coupler for phase shifting.

The modulator in Fig. 10 represents an XOR/XNOR logic function. The potential V is fixed on the lower branch such that a phase shift of π is induced into that branch. A potential applied at either V_1 or V_2 also induces a phase shift of π into the upper branch. The net phase shift upon recombination is $\pi - \pi = 0$, and the output power is high. If, however, V_1 and V_2 are both high or low simultaneously, the net phase shift is π, and the output optical power is zero. Reverse outputs for these input conditions (XNOR) can be achieved if the voltage V on the lower branch is removed.

If a fully optical logic gate is to be realized in a device similar to Fig. 10, then optical detectors must be used to convert incoming optical logic into the electrical equivalents V_1, V_2, and V. This involves mounting hybrid photodetectors on the ends of the input waveguides.

Figure 14 shows a D/A converter realized in an electro-optic waveguide. The device takes advantage of the fact that the electro-optic phase shift given by Eqs 54, 55 is also dependent on the length of the electrode. Applying a signal to the least significant bit results in a smaller phase shift as applying signal to the most significant bit.

2.4 TIR Switches

Another type of electro-optic switch is based on control of the refractive index during total internal reflectance (TIR) at the intersection of two waveguides. An example of a TIR switch is illustrated in Fig. 15. The figure illustrates a device based on the refractive index variation of an electro-optic material placed at the intersection of two guides. This index variation may be positive or negative. Light is incident to the switch from

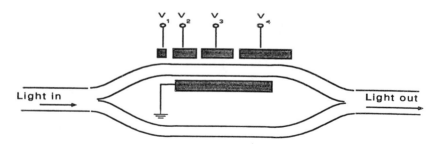

Figure 14 Digital/analog converter realized in a electro-optic waveguide.

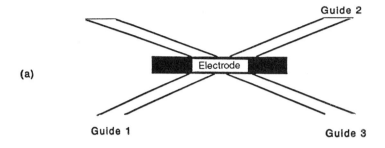

(a)

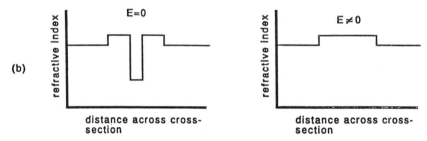

(b)

Figure 15 Optical switch based on total internal reflection (TIR). The electrode in the center of the switch is fabricated in electro-optic medium, so the index of refraction at the intersection can be either increased or decreased, depending on whether a positive- or negative-index switch is desired.

waveguide 1. For a positive-index device, when the switch is off the external electric field is zero, and the refractive index in waveguide 1 is such that the light simply passes from waveguide 1 to waveguide 2.

When an electric field is placed on the electrode at the center of switch, the refractive index increases for the positive-index switch. Figure 15b illustrates the index profile through a section at the center of the guide for a positive index switch. Under an applied field, the incident angle of waveguide 1 to the switch is greater than the critical angle for total internal reflection, and the light from waveguide 1 is reflected into waveguide 3.

For a negative-index switch the refractive index at the center is higher when no external field is present. Light is then diffracted into waveguide 3 when the switch is off. The application of the field acts to lower the index at the center of the switch, the TIR condition is destroyed, and the light passes through to waveguide 2.

The TIR switch has several advantages over other electro-optic switches, including small size, operable over a long range of wavelengths, and polarization independence. Since the critical angle for TIR depends on the ratio of the refractive index differences, a large electro-optic coefficient is desirable to allow for a large intersecting angle and narrower waveguides.

As with other devices based on the electro-optic effect, operation of the TIR switch depends on the value of the electro-optic coefficient. For a compound semiconductor this coefficient is larger for wavelengths of light corresponding to energies close to the band-gap energy. However, the value of the absorption also increases for photons of energy close to the band-gap energy. Since absorption is related to the imaginary part of the dielectric constant, Shimomura et al. [6] have introduced an "index loss variation ratio, which is defined as the ratio of the real part of the variation of dielectric constant ($\Delta n'$) to the variation of the imaginary part ($\Delta n''$):

$$\overline{\alpha}_p = \frac{\Delta n'}{\Delta n''} = \frac{4\pi n_{eq}}{\lambda} \frac{\Delta_{eq}}{\Delta \alpha_{loss}} \tag{97}$$

where λ is the wavelength and Δ_{eq} and $\Delta \alpha_{loss}$ are variations in the refractive index and absorption coefficient, respectively, of the waveguide material. Shimomura calculates that for compound semiconductor TIR devices, a value of $\alpha_p > 10$ is needed in order to minimize the loss in the waveguide. The value of α_p is higher for positive- than for negative-index switches fabricated using compound semiconductors. If a negative-index switch is used in compound semiconductor waveguides, total internal reflection is destroyed by the increase in optical absorption, and such a large value of α_p is difficult to obtain. A positive-index change is then necessary to provide minimal insertion loss for the device. A negative-index change is desirable, however, as a negative change is usually almost twice the size of a positive change for the same electric field.

Shimomura also pointed out that a large value of α_p could be obtained with low-dimensional structures, such as quantum wells, wires, or quantum dots. Figure 16 shows the calculated index variations expected for $Ga_x In_{1-x} As/InP$ quantum box, wire, and film structures. Due to the decrease in the dimensionality, the wavelength for band-to-band absorption is less as the carriers become restrained in each dimension. (Refer to Chapter 1, Section 14.) Hence, for the device in Fig. 16, the absorption wavelengths for quantum box, wire, and film are 1.407, 1.506, 1.584 μm, respectively. Figure 17 shows the values of the index loss variation, α_p,

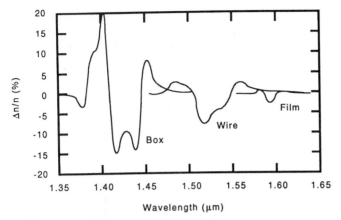

Figure 16 Refractive index variation for GaInAs-InP quantum film, wire, and box structures, as calculated by Shimomura [6]. For the calculations, dimensions of quantum film, wire, and box were taken as 10 nm, 10 nm × 10 nm, and 10 nm × 10 nm × 10 nm, respectively. The applied electric field was taken at 1×10^5 V/cm.

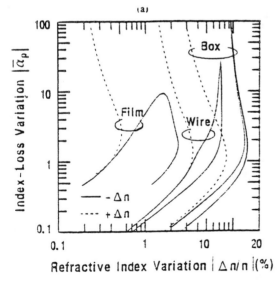

Figure 17 Index loss coefficient α_p vs. refractive index change for GaInAs-InP quantum film, wire, and box structures of Fig. 16 [6]. Solid lines represent negative-index variations, while dotted lines represent positive-index variations.

for these structures. Solid lines represent where $\Delta n < 0$ and dotted lines where $\Delta n > 0$. The absolute value of the index change for the negative-index variation is larger than that from the positive-index variation. For the negative-index devices, no values of $\alpha_p > 10$ are possible for quantum film devices. Because of the change of density of states from staircase to sawtooth to delta function for the change from film to wire to box, the value of the effective index loss is raised for these lower-dimensional structures. Thus it is desirable to prepare devices in such structures.

2.5 Y-Branch Modulator/Switch

Figure 18A demonstrates schematically the operating principles of the Y-branch modulator/switch. Two electrodes are fabricated on each side of a branching Y formed from a central optical waveguide. The waveguide is designed for single-mode operation. The field distribution at the branching point of the waveguide can be altered via the electro-optic effect by applied

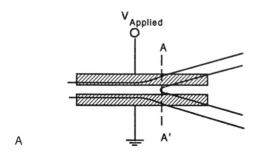

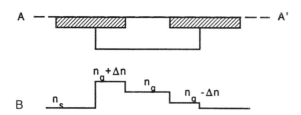

Figure 18 (A) A Y-branch optical switch/modulator formed from a single-mode central waveguide. (B) The distribution of the refractive index at the branching point of the waveguide in (A) under an applied electric field. The applied voltage forces an asymmetry in the guide which channels light away from the lower index arm into the higher-index arm.

voltages placed on the electrodes. With no applied voltage, the distribution of the refractive index is the same throughout the central guide as well as in each branching arm, and each branch arm guides an equal amount of light away from the central guide. Application of an electric field alters the refractive index at the branching point as shown in Fig. 18B. The applied voltage forces an asymmetry in the refractive index distribution at the branching point. The optical power is forced down the branching arm with the higher refractive index and channeled away from the arm with the lower index. The switch can be reversed by reversing the polarity on the electrodes. The asymmetry in the refractive index is then reversed, and the light is channeled down the other arm.

The Y-branch modulator is not an interferometric device and is easier to fabricate and control than the Mach-Zehnder modulator/switch. The branching point of the Y must be designed to prevent multiple orders from being excited at that point; otherwise the device is more tolerant of variations in geometry and voltage than the interferometric device and is suitable for switching matrixes and multiplexors.

3 ELECTROABSORPTION MODULATORS

Although electroabsorption modulators are not based on Pockel's effect, they still qualify as electro-optic modulators in the sense that an electric field controls the optical switching mechanism. The electro-absorption device is based on the Franz-Keldysh effect, which was commented on in Section 11.3 of Chapter 1. As demonstrated in Fig. 14 in Chapter 1, a high electric field placed on a semiconductor device causes the energy bands of the semiconductor to bend, and the exponential tails of the electron wavefunctions then extend further into the energy gap. The semiconductor can then absorb photons of a lower energy, and hence a longer wavelength, than it can when the high electric field is removed. The absorption edge of Fig. 15 in Chapter 1 is effectively shifted to longer wavelengths as the electric field strengthens. The effective change in band-gap energy, ΔE, was shown in Eq. 54 of Chapter 1 to be related to the applied electric field $\mathscr{E}$ by the relation

$$\Delta E = \frac{3}{2} (m^*)^{-1/3} (q\hbar\mathscr{E})^{2/3} \tag{98}$$

where m^* is the effective mass of the electron. Since the absorption coefficient is a strong function of electric field, the absorption can be greatly altered. For a direct-band-gap semiconductor such as GaAs, an electric field of 10^5 V/cm can increase the absorption coefficient by almost three

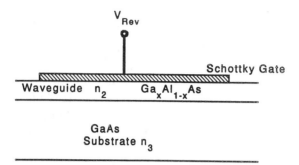

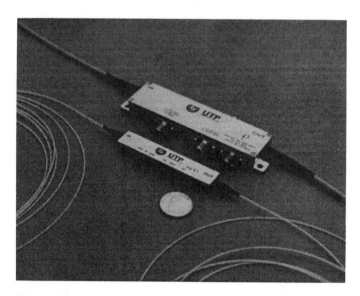

Figure 19 Optical modulator taking advantage of the electroabsorption (Franz-Keldysh) effect. The composition of Ga$_x$Al$_{1-x}$As is chosen such that the band-gap energy is slightly larger than the photon energy. Application of an electric field induces absorption at the photon energy by lowering the effective band-gap energy.

Figure 20 Commercial 2.5 GB/s digital electro-optic waveguide modulators, shown in SMA and butterfly packages, are fabricated on lithium niobate using an ion-exchange process. (Photo courtesy of Uniphase Telecommunication products).

orders of magnitude for photons of energy slightly less than the zero-field band-gap energy.

An electroabsorption device can then be fabricated to switch photons of energy slightly less than the band-gap energy. Such an electroabsorption device is shown in Fig. 19. A nonuniform electric field is present in a junction, such as a Schottky barrier or a *p-n* junction. The electric field at the surface of the Schottky junction in Fig. 19 is high enough to promote absorption via the Franz-Keldysh effect. The composition of $Al_bGa_{1-b}As$ in the waveguide is chosen to provide a tailored band-gap energy slightly higher than the photon energy. The doping in the substrate is chosen to be higher than that in the guide, so that the electric field is concentrated mainly in the guide. Such a modulator has been demonstrated to be effective in controlling the optical intensity though electric-field-dependent absorption.

REFERENCES

1. S. Blakemore, Semiconducting and other major properties of GaAs, *J. Appl. Phys.*, *53*, R123–R181 (1982).
2. A. Yariv, Coupled mode theory for guided wave optics, *IEEE J. Quantum Electron.*, *QE-9*, 919 (1973).
3. A. B. Buckman, *Guided Wave Photonics*, Saunders, 1992, Chapter 5.
4. C. H. Bulmer, W. K. Burns, and R. P. Moeller, Linear interferometric waveguide modulator for electromagnetic-field detection, *Opt. Lett.*, *5*, 176 (1980).
5. M. Minakata, Efficient LiNbO₃ balanced bridge modulator/switch with an ion-etched slot, *Appl. Phys. Lett.*, *35*, 40–42 (1979).
6. K. Shimomura, Y. Suematsu, and S. Arai, Analysis of semiconductor intersectional waveguide optical switch/modulator, *IEEE J. Quantum Electron.*, *QE-26*, 883 (1990). K. Shimomura, S. Arai, and Y. Suematsu, Operational wavelength range of GaInAs(P)-InP Intersectional optical switches using field-induced electrooptic effect in low-dimensional quantum-well structures, *IEEE J. Quantum Electron.*, *QE-28*, 471 (1992).

BIBLIOGRAPHY

H. Kogelnik, Theory of dielectric waveguides, in *Integrated Optics*, 2nd ed. (T. Tamir, ed.), Springer-Verlag, Berlin, 1979, p. 13.

M. Papuchon, LiNbO₃ devices, in *Waveguide Optoelectronics* (J. H. Marsh and R. M. De La Rue, eds.), Kluwer Academic, 1992, p. 73.

R. G. Hunsperger, *Integrated Optics: Theory and Technology*, 2nd ed. Springer-Verlag, Berlin, 1985.

H. Nishihara, M. Haruna, and T. Suhara, *Optical Integrated Circuits*, Macmillan, New York, 1987, 5 p. 96.

Ka-Kha Wong, Ed *Integrated Optical Circuits*, Boston, MA, Proceedings SPIE, Vol. 1583, 1991.

T. Sueta and M. Izutsu, High speed guided-wave optical modulators, *J. Opt. Commun.*, *3*(2), 52–58 (1982).

R. Syms and J. Cozens, *Optical Guided Waves and Devices*, McGraw-Hill, London, 1992.

J. A. del Alamo and C. C. Eugster, Electron waveguide devices, in *Gallium Arsenide and Related Compounds 1992*. Proceedings of the Nineteenth International Symposium (T. Ikegami. F. Hasegawa, and Y. Takeda, eds.) Karuizawa, Japan, 1993.

M. Kondo, LiNbO$_3$/waveguide devices, *J. Ceram. Soc. Jpn.*, *101*(1), 38–42 (1993).

O. Wada, III-V semiconductor integrated optoelectronics for optical computing Proc. SPIE—Int. Soc. Opt. Eng., vol. 1362, pt. 2, pp. 598–607, 1991. *Physical Concepts of Materials for Novel Optoelectronic Device Applications II: Device Physics and Applications*. SPIE. Aachen, Germany, 1991.

9

Liquid Crystal Devices

1 INTRODUCTION

When certain organic substances are heated from solid to liquid, they go through an intermediate phase, or even a succession of intermediate phases, before completely melting. Such phases are called *mesophases* and are usually characterized as being milky or slightly opaque in appearance. Mesophases are subdivided into two types: (1) disordered mesophases, or plastic crystals, are spherical molecules; (2) ordered mesophases, or liquid crystals, are rod-shaped molecules.

In a liquid crystal the rodlike molecules are arranged in an ordered fashion, but the chemical bonding between the rods is too weak to sustain the more familiar solidlike structure. There is long-range order in the structure, but the molecules are able to "slide" past one another and thus form a structure much weaker than the crystalline phase. Liquid crystal molecules generally have a length of ~2.5 nm and a diameter of ~0.5 nm, with a molecular weight of 200–500 g/mole. To form liquid crystals the molecules must have a rigid backbone, and they are not articulated like —CH_2— chain molecules.

Liquid crystals can be further subdivided into *lyotropic* and *thermotropic*. Lyotropic liquid crystals are generally biological molecules (such as DNA) dissolved in a solvent. Because of the solvent, lyotropic liquid crystals tend to decompose in an applied electric field and therefore are not generally useful in commercial optoelectronic technologies.

Thermotropic liquid crystals are used in commercial optoelectronic devices and, as the name suggests, are strongly susceptive to any fluctuations in temperature. A rapid rise in the temperature can seriously weaken the loose bonds between the molecules. Early examples of thermotropic liquid crystals include derivatives of cholesteryl and the substance MBBA (see Fig. 1).

There are three main types of rod-shaped thermotropic liquid crystals illustrated in Fig. 2:

Nematic crystals (Fig. 2a). In this state the rodlike molecules are oriented parallel to one another, but their longitudinal positions are randomly placed.

Smectic crystals (Fig. 2b). The rods are parallel, as in the nematic state, but there is also symmetry in the longitudinal direction of the molecules. The molecules are packed in parallel layers forming molecular "planes" in the longitudinal dimension.

Cholesteric crystals (Fig. 2c). The rods are ordered in parallel layers, but the molecules tend to "twist" in the third dimension, thus giving a helical smectic state. The name comes from the fact that many of the earlier cholesteric crystals were derivatives of cholesteryl, but now include several other types of chiral nematic crystals.

Some substances pass from crystalline solid to smectic or nematic to liquid as the substance is heated. Others pass through several phases, for example, from crystalline, to smectic, to nematic, to isotropic liquid.

In addition, each of the above liquid crystal phases may be further broken into relative phases. *Homeotropic* ordering of a nematic liquid crystal is shown in Fig. 3a, where the rods are oriented perpendicular to the surface of an interface. Where the rods are ordered parallel to the interface, as in Fig. 3b the ordering is said to be *homogeneous*.

Nematic liquid crystals can change from homogeneous to homeotropic upon application of a suitably oriented electric field, for example, if the "interfaces" of Fig. 3 are electrodes. This is because the dielectric con-

Figure 1   Molecular structure of MBBA (short of *N*-(*P*-methoxy benzylidene)-*P'*-butylaniline) which was the first example of a single Schiff base compound to exhibit nematic behavior at ambient temperature.

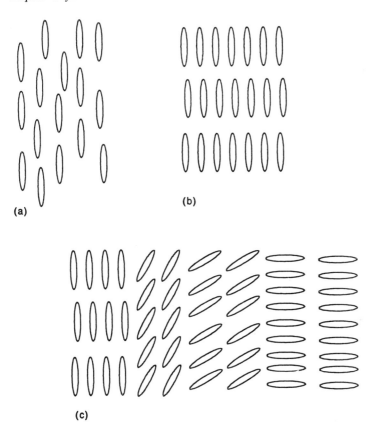

(a)

(b)

(c)

Figure 2 Schematic of the difference in ordering of rodlike liquid crystal molecules. (a) In the nematic state, the rods are parallel but lack any longitudinal ordering. (b) In the semectic state, the rods have both lateral and longitudinal symmetry. (c) In the cholesteric state the symmetry is similar to the semectic state, however, the plane of rodlike molecules rotates in the third dimension.

stant of the liquid crystal with the field parallel to the molecules ($\epsilon_\parallel$) is different from that with the field perpendicular to the rods ($\epsilon_\perp$). Since there is a difference in the dielectric constant (and therefore also in the refractive index), with position in the liquid crystal, the crystal is an electro-optic material, and has properties entirely similar to the electro-optic crystals discussed in Chapter 8. If $\epsilon_\parallel \gg \epsilon_\perp$, then the liquid crystal is said to be a "positive" electro-optic with the optical axis parallel to the direction of the rods. The application of an electric field to the electrodes will cause the molecules to rotate in order to minimize their electrostatic

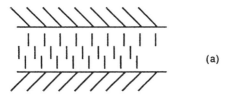

(a)

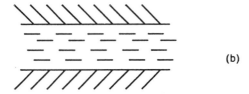

(b)

Figure 3   (a) Homeotropic and (b) homogeneous ordering of a nematic liquid crystal.

energy, until the longitudinal axis of the molecules is in the direction of the applied field. This transition of molecular alignment is found to take place above a critical electric field, $\mathscr{E}_c$. The exact quantity, as well as the sign of the dielectric anisotropy, depends upon the chemical structure of the liquid crystal molecule. Different chemical side groups and polarization constituents to the molecule affect dielectric and optical properties. For example, the presence of cyano (—C—N=O) groups on the molecules can greatly increase the dielectric anisotropy. Aromatic groups (such as benzene) also affect the anisotropy, because of the polar nature they contribute through their π-bonding electrons.

A transparent liquid crystal may become milky white once an electric field is applied. This is due to the increase in scattering centers within the liquid, once the molecules align with the field. These scattering centers are caused by the formation of molecular "domains," which are microscopic grouping of molecules. The presence of the surface (at the electrode) serves to make the liquid crystal a single domain. If the bounding surfaces have not been specially treated to orient the molecules with respect to the surface, the liquid crystal will have several domains and the molecules will have several inclinations in orientation. Some liquid crystal devices make use of the dynamic scattering of light from unoriented domains. Alignment is such that the liquid crystal is viewed perpendicular to the electrodes, and at least one or more electrode must be transparent. For most commercial devices, however, the unbiased state is characterized by the alignment of the rod molecules with respect to one or both electrodes.

2 TWISTED NEMATIC CELLS

Of the three liquid crystal phases, most commercial devices are constructed from nematic materials. Nematics have the least amount of molecular ordering, have the smallest molecular steric constraints, and have the lowest viscosities. Thus they respond most easily to an applied electric field.

The simplest type of liquid crystal device is a "twisted nematic" device, where a nematic phase liquid crystal is prepared as in Fig. 4. The surfaces of a transparent substrate are coated with a transparent conductor such as indium tin oxide (ITO). For a seven-segment display the electrode is patterned appropriately. Two of these substrates are positioned 6 to 8 microns apart, with liquid crystal material filled within. The interior surfaces of the substrate are treated with a material and/or polished is such a way that the rodlike molecules of the liquid crystal (in a homogeneous state) tend to orient in a preferred direction along the surfaces of the electrodes. For example, a layer of polyimid coated on the ITO layer, polished with parallel strokes, can be used to orient the molecules.

The cells may operate in the transmission mode, as shown in Fig. 4a, or in the reflection mode. The orientation of the molecules along one surface is twisted 90° with respect to the opposite electrode, to achieve a twisted state to the molecular planes along the z-direction. The optic axis of the liquid crystal is parallel to the x-y plane and rotates with the z-direction. One can envision this as a rotation of several stacked index ellipsoids of Fig. 3 of Chapter 8 stacked in the z-direction. Each subsequent ellipsoid is rotated with respect to the previous one. Application of an electric field to the electrodes will cause the molecules to align themselves along the direction of that field, as shown in Fig. 4b, and the optical axis of the liquid crystal rotates as well until it is parallel to the z-plane.

Along the axis of the twist in the liquid crystal planes with no electric field, the twist angle ϕ is linear with distance z:

$$\phi = \alpha z \tag{1}$$

where α is a proportionality constant defined as the *twist coefficient*. The amount of phase retardation per unit length is defined as

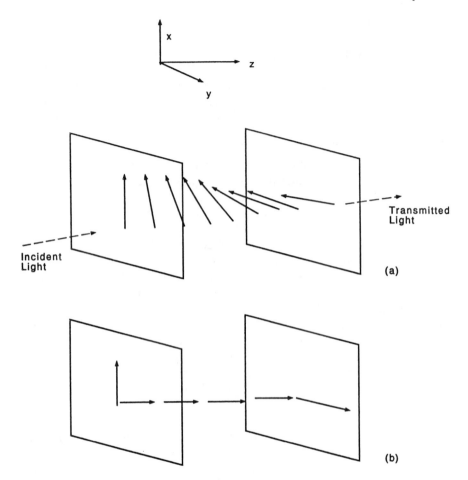

Figure 4 The twisted nemetic cell. In (a), under no applied electric field, the liquid crystal is arranged with the molecules aligned parallel to the surface of the electrodes, in the homogeneous order. The alignment direction of one electrode is rotated 90° with respect to the other, causing a "twist" in the molecular orientation along the length of the cell. In (b), under an electric field, the molecules align with the direction of the field.

$$\beta = \Delta nk \tag{2}$$

$$\Delta n = n_e - n_o \tag{3}$$

where n_o and n_o are the extraordinary and ordinary refractive indices of the crystal, and k is the wavevector. The electric field vector of plane polarized light incident on the end of the twisted cell will decompose into two components, one traveling at a velocity c/n_o and the other at c/n_e. Since the optic axis rotates with the dimension z according to Eq. 1, the magnitude of each of the two components varies along the length of the cell. At the end of the cell, the optical axis has rotated by 90°, and the components of the electric field have rotated by that same amount. The resultant of those components also has rotated 90°, and the exiting plane of polarization of the light has rotated that amount with respect to the incident plane. The liquid crystal then serves as a polarization rotator for plane polarized light. The wave maintains the plane polarized state, but the plane of polarization rotates along with the twist angle α. The phase retardation simply supplies a phase shift given by βd, where d is the thickness of the liquid crystal.

The liquid crystal cell can be used as a phase modulator, with the amount of retardation given by βd. If cross-polarizers are placed at each end of the cell, as shown in Fig. 5a the device acts as a voltage-controlled intensity modulator. Light is incident on the first polarizer, and when the voltage is removed the liquid crystal rotates the plane of polarization through the cell and out the exit polarizer. When the voltage is applied to the cell, the plane of polarization is no longer rotated and the light is blocked by the exit polarizer. Thus the device operates much like the bulk Pockels cell device.

A more common LCD display operates in reflection mode. In Fig. 5b, light entering the device is initially plane polarized by an entrance/exit polarizer. When the device is off, the plane of polarization of the light is rotated by the twisted nematic cell as it passes through the cell. The light then reflects off a reflector mounted behind the cell and is rotated back through the entrance/exit polarizer. To the eye, the cell then appears bright. If a voltage is applied, the light can no longer reach the rear reflector and the cell appears black. The cells can easily be fabricated in an addressable matrix with crossed row/column electrodes. Since it does not have to produce light, the device consumes little power. A typical LCD uses a few microwatts per square centimeter of display area, which is only a few thousandths of a percent of what a light-emitting display requires.

The device is normally operated in AC mode using a square wave, from a few tens of hertz to several kHz. Operating the cell at lower frequencies causes electromechanical decomposition of the cell molecules. Devices are usually driven with AC signals and with no DC component.

When a voltage V is applied to a cell of thickness d such that $V > V_c$, where V_c is the critical voltage, the molecules rotate in the direction of the field at an angle θ to the substrate, called the "tilt" angle. The value of the tilt angle varies from near zero at the substrate edges to a maximum angle at the mid-layer of the liquid crystal. The amount of tilt to the molecules at mid-layer can be described by [1]

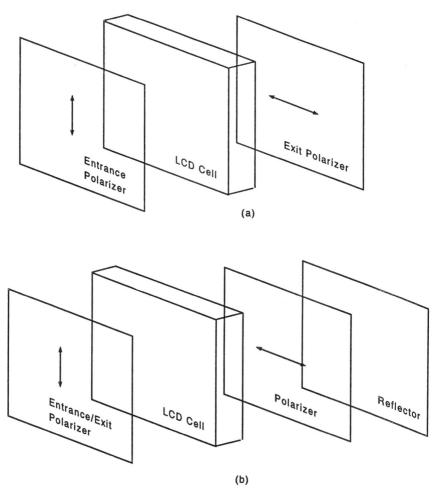

(a)

(b)

Figure 5 Twisted nematic cell operating in (a) transmission and (b) reflection.

$$\theta = \frac{\pi}{2} - 2 \tan^{-1} \exp\!\left(\frac{-(V - V_c)}{V_0}\right) \tag{4}$$

where V_0 is a constant. Experimentally, a value of θ of 80° usually requires an applied voltage of roughly three times the critical voltage. For light traveling through the cell the index of refraction $n(\theta)$ is given by the birefringence:

$$\frac{1}{n^2(\theta)} = \frac{\cos^2 \theta}{n_e^2} + \frac{\sin^2 \theta}{n_o^2} \tag{5}$$

and the light undergoes a phase retardation given by

$$\Gamma = \frac{2\pi d}{\lambda} [n(\theta) - n_o] \tag{6}$$

A plot of the phase retardation with applied voltage is shown in Fig. 6. The transition from off to on appears at a value of the applied voltage which is roughly 1.5 times the critical voltage as measured from where it initially breaks threshold. Values of V_c usually range between 0.5 and 2 V_{rms}, with a value of 2.5 to 3 times V_c necessary to rotate the plane of polarization by a value of $\pi/2$.

2.1 Supertwisted Nematic LCDs

Nematic liquid crystals with a "supertwist" have been developed in an effort to make the curves such as that of Fig. 6 sharper.

For a twisted nematic to actually rotate the plane of polarization by 90° requires a value of β in Eq. 2 such that

$$\beta d \gg 1 \tag{7}$$

which follows the requirement $\beta \gg \alpha$. Thus the twist in the nematic crystals must be gradual, during which the light must undergo several degrees

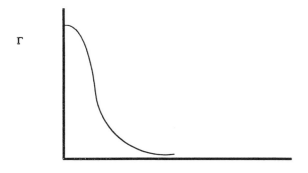

Figure 6 Phase retardation vs. applied voltage for a liquid crystal cell. The steeper the Γ vs. voltage curve, the sharper the critical voltage for the cell.

of phase shift. Values of $n_e - n_o$ are typically between 0.1 and 0.2. This means that a relatively thick liquid crystal must be used in order to satisfy this condition. Thick layers of liquid crystal are undesirable since the speed of the device is determined by the rate of rotation of the molecule and, hence, by how much molecule is present in the device. However, it was found that if the twist in the unbiased molecules is increased from 90° to more than 180°, the steepness of the curve in Fig. 6 would increase to almost vertical at a pretwist angle near 270° due to an increase in phase shift per unit length. In addition, some of the supertwist nematic devices will have a "pretilt" angle to the molecules. The pretilt angle implies that the molecules close to the substrate are tilted slightly under no bias. This pretilt also makes the threshold voltage better defined and lowers the threshold. Pretilt is given to the molecules by adding a dopant to the liquid crystal, which causes the molecules to become chiral and tilt with respect to the substrate.

2.2 Optical Rotary Power in Twisted Nematics

The conditions for rotation of the polarization vector of plane polarized light by a twisted electro-optic has been demonstrated by de Vries [2]. First, the cell thickness is divided into several incremental layers of equal width. Each layer consists of a positive electro-optic, with ϵ_1 and ϵ_2 being the dielectric constants of the fast and slow directions, respectively (that is, $\epsilon_2 > \epsilon_1$). The x- and y-axes are laid along the first layer, with the x-axis parallel to ϵ_1 and the y-axis parallel to ϵ_2. For each successive layer the axis has turned through an angle ϕ, as shown in Fig. 7, as the privileged directions also rotate through ϕ. In the second layer the coordinate system is now the ξ-η coordinate system. The light is incident along the z-direc-

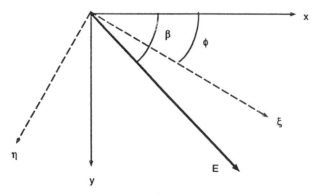

Figure 7 Coordinate system used by deVries [2] to demonstrate optical rotary power in a twisted liquid crystal cell.

tion, and upon passing from one layer to the next experiences a change in the refractive index. The coordinates of the electric field vector, E_x and E_y, have changed to those of the new coordinate system, E_ξ and E_η.

Along the z-axis, the twist angle ϕ is

$$\phi = \alpha z \tag{1}$$

where α is the twist coefficient. One can relate the x and y coordinates of the electric field vector to those of the ξ-η coordinate system by

$$E_x = E_\xi \cos \alpha z - E_\eta \sin \alpha z \tag{8a}$$

and

$$E_y = E_\xi \sin \alpha z + E_\eta \cos \alpha z \tag{8b}$$

The wave equation along the x and y directions can be written as

$$\frac{\partial^2 E_x}{\partial z^2} - \frac{\epsilon_1}{c^2} \frac{\partial^2 E_x}{\partial t^2} = 0 \tag{9a}$$

$$\frac{\partial^2 E_y}{\partial z^2} - \frac{\epsilon_2}{c^2} \frac{\partial^2 E_y}{\partial t^2} = 0 \tag{9b}$$

Substituting Eqs. 8 into Eqs. 9 yield two superimposed plane waves. Equating the amplitudes of each gives us

$$\frac{\epsilon_1}{c^2} \frac{\partial^2 E_\xi}{\partial t^2} = \frac{\partial^2 E_\xi}{\partial z^2} - 2\alpha \frac{\partial E_\eta}{\partial z} - \alpha^2 E_\xi \tag{10a}$$

and

$$\frac{\epsilon_2}{c^2} \frac{\partial^2 E_\eta}{\partial t^2} = \frac{\partial^2 E_\eta}{\partial z^2} + 2\alpha \frac{\partial E_\xi}{\partial z} - \alpha^2 E_\eta \tag{10b}$$

Solutions for Eqs. 10 are given by

$$E_\xi = A \exp[kmz - \omega t) \tag{11a}$$

$$E_\eta = iB \exp[kmz - \omega t) \tag{11b}$$

which represent an elliptically polarized wave. Substituting Eqs. 11 into Eqs. 10 gives

$$Ak^2\epsilon_1 = Am^2k^2 + 2Bmk\alpha + \alpha^2 A \tag{12a}$$

$$Bk^2\epsilon_2 = Bm^2k^2 + 2Amk\alpha + \alpha^2 B \tag{12b}$$

Solving for B/A and then for m^2, we obtain

$$m^4 - m^2 \left(\epsilon_1 + \epsilon_2 + \frac{2\alpha^2}{k^2} \right) + \left(\epsilon_1 - \frac{\alpha^2}{k^2} \right)\left(\epsilon_2 - \frac{\alpha^2}{k^2} \right) = 0 \tag{13}$$

There are two roots for m^2 (m_1^2 and m_2^2) in Eq. 13. Each root corresponds to a wave with an elliptical polarization given by the ratio of B/A, which can be determined from Eqs. 12. The ratio can be found from solving for B_1/A_1 and A_2/B_2 using values of m_1 and m_2. In the *Mauguin condition*—that is for

$$\epsilon_1 - \epsilon_2 \gg \left(\frac{2\alpha}{k}\right)^2 \tag{14}$$

—the values of ϵ_1 and ϵ_2 reduce to

$$\epsilon_1 = \sqrt{m_1} \tag{15}$$

and

$$\epsilon_2 = \sqrt{m_2} \tag{16}$$

and the m's are simply the values of the birefringence (the values of the refractive indices for the e-ray and o-ray). The ratios of B/A and A/B can be determined to be

$$\frac{B_1}{A_1} = \frac{k^2\epsilon_1 - m_1^2 k^2 - \alpha^2}{2m_1 k\alpha} \approx 0 \tag{17}$$

$$\frac{A_2}{B_2} = \frac{k^2\epsilon_2 - m_2^2 k^2 - \alpha^2}{2m_2 k\alpha} \approx 0 \tag{18}$$

which means the two normal waves are linearly polarized with amplitudes A_1 and B_2. The first wave is polarized along the ξ-axis, the second along the η-axis. According to Eqs. 11, E_η and E_ξ will be out of phase. Their sum is a linear polarized wave of equal magnitude to the original plane wave but rotated with respect to its position in the x-y coordinate system. This corresponds to a rotation of the plane of polarization of the light, and the rotation corresponds to the twist angle of the liquid crystal molecules.

The Mauguin condition can also be written as

$$p\Delta n \gg 2\lambda \tag{19}$$

where p is the pitch and Δn is the birefringence ($m_2 - m_1$). For a twisted nematic of $\pi/2$ twist angle, this becomes

$$d\Delta n \gg \frac{\lambda}{2} \tag{20}$$

where d is the cell thickness. Equation 20 places a minimum value on the thickness of the cell. Gooch and Tarry [3] have shown that for ideal crossed polarizers, the ratio of output transmitted light to incident is given

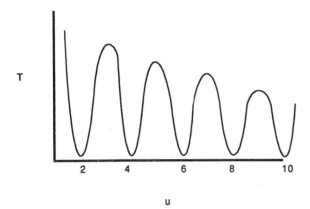

Figure 8 An example of a Gooch-Tarry curve. The plot is transmitted light from an LCD cell with crossed polarizers vs. the parameter u, which is defined in the text.

by

$$T = \frac{\sin^2[(\pi/2)\,(1 + u^2)^{1/2}]}{1 + u^2} \tag{21}$$

where

$$u = 2d\,\frac{\Delta n}{\lambda} \tag{22}$$

a plot of which is shown in Fig. 8. In order to maximize contrast of the cell, a value of u must be chosen to minimize the value of T. Since the value of Δn is fixed by the liquid crystal medium, the varying parameter is d. The thickness of the cell can be fixed by placing small spacers between the electrode plates. A small value of u means a higher switching speed for the device and a larger viewing angle, since the light has less distance to travel through the cell. However, this is at the expense of a lower yield for the cell, since high precision in the value of d makes very thin cells more difficult to process.

3 FERROELECTRIC LIQUID CRYSTALS

The smectic state, which was illustrated in Fig. 2, is characterized by ordered rows of molecules, each row existing in a single plane and each plane stacked upon on another as in sheets or plates. The smectic state of liquid crystals can be subdivided into 12 individual classes, which are assigned letter designations and differ only in the positional ordering and relative tilt of the molecules. For seven of these classes the molecules are tilted at an angle with respect to the stacking planes. These seven phases are known as C, F, G, H, I, J, and K. The molecules of each of these phases are chiral, and the chiral molecules are designated C^*, F^*, G^*, H^*, I^*, J^*, and K^*.

An example of a chiral molecule is shown in Fig. 9. The molecule exists in two forms, one is a mirror image of the other. For liquid crystals, only three of the chiral phases, C^*, I^*, and F^*, form smectic phases. These three phases possess a unique ordering along the direction of the stacking planes. As one passes from plane to plane in one of these phases, the tilt angle of the molecules varies slightly. This precession of tilt orientation, when viewed over several layers, forms a helical structure. The structure of a ferroelectric liquid crystal is shown schematically in Fig. 10. The helix is a macroscopic ordering of the molecules and *in itself* is of a chiral nature. Thus, there exist left-handed helices as well as right-handed versions. The helix has the property of rotating plane polarized light which is transmitted along a direction parallel to the axis of the helix. This macroscopic property gives these phases ferroelectric properties; that is, the electrical anisotropy of the liquid crystal is different depending upon the orientation of the molecules.

If a ferroelectric liquid crystal is placed between two transparent electrodes, only two stable states are possible. The liquid crystal molecules first twist to make a tilt angle of θ with respect to the electrode surface, with one stable state for a tilt angle of $+\theta$, and one for $-\theta$. The axis of the macroscopic helix lies in the direction of the electric field, and the electric field allows the molecules to reorient from one tilt angle to the other. The cell then acts as an electro-optic crystal whose optical axis can be switched between two different orientations. The optical axis lies along the direction given by angle $\pm\theta$.

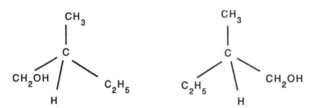

Figure 9   An example of a chiral molecule. The molecule has two forms, one of which is the mirror image of the other.

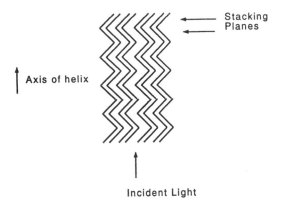

Incident Light

Figure 10 Macroscopic helical ordering of the chiral phases of a smectic liquid crystal. The helical ordering is formed due to a twist angle at each successive molecular plane with respect to the plane below and above.

Consider plane polarized light incident on such a cell with the plane of polarization at an angle of $+\theta$ with respect to the surface. When the liquid crystal is in the $+\theta$ state, the plane of polarization of the incident light lies along the optical axis of the liquid crystal. In the $+\theta$ state there is only one component of the electric field vector of the light, which travels in the medium with refractive index n_e. If an electric field is applied in such a way that the liquid crystal changes to the $-\theta$ state, the plane of polarization now makes an angle of 2θ with respect to the optic axis. If $2\theta = 45°$, the light travels with two components, n_e and n_o. The phase retardation is given by

$$\Gamma = 2\pi \frac{d}{\lambda} (n_e - n_o) \tag{23}$$

and if d is chosen such that $\Gamma = 90°$, the plane of polarization of the light will rotate 90°. A modulator, similar to the twisted nematic device, can then be constructed if the cell is placed between crossed polarizers.

The large number and variety of phases of smectic crystals and the large number of chiral phases mean that a large variety of ferroelectric liquid crystal devices of different properties can be formed. Some phases possess very high switching speeds compared to nematic liquid crystals. Others have slow speeds, but very sharp thresholds, making them useful as storage devices.

4 COLOR LCDS

Color can be added to liquid crystal displays by several methods. Color LCDs can be made by adding the appropriate color filter to the transparent substrate. Color filters can be added in several different geometries, most involving color stripes of red, blue, and green mounted over transparent liquid crystal displays with backlights. Light emitting from the various color filters can be blocked out by applying a voltage to the appropriate liquid crystals. Alternatively, color polarizers can be used in conjunction with the liquid crystal, and different colors can be produced by control of the polarization of the light exiting the LCD pixels. Three multiplexed displays are then placed in parallel, and each pixel is capable of a three-color output through subtractive color filtering.

LCD devices also have the capability of producing their own colors naturally. If the value of Δn is large, a nematic will exhibit a large birefringence. If the nematic cells are made thin enough, and the light is viewed between crossed polarizers, interference colors will be generated. The color depends upon the film thickness and the amount of birefringence. The molecular alignment, and therefore the color, can be changed by application of an electric field. Under an electric field, stacks of homogeneous nematic cells with positive anisotropy will align perpendicular to the field while heterogeneous cells with negative anisotropy will align parallel to the field. Either cell will exhibit large birefringence and a change of color on reordering.

Cholesteric liquid crystals, as was demonstrated in Fig. 2, exist in layers stacked in a helical fashion. Incident light can Bragg-reflect off of the constructive interference created by the stacked layers, similar to the way light reflects off of refractive index gratings in acousto-optic devices (see Section 2 of Chapter 7). The color of the light diffracted depends upon the average spacing between the cholesteric domains, which in turn depends upon the helical pitch of the molecules. This helical pitch is sensitive to temperature, ultraviolet irradiation, and the presence of certain organics, and is therefore most useful in visible sensor applications. Predominant are visible liquid crystal cholesteric thermometers.

Unfortunately, both cholesteric color crystals and nematic birefringent color liquid crystals are not suitable for most color display technology, mainly due to the sensitivity of effects on temperature, applied voltage, and variations in film thickness. Most commercial color LCDs have concentrated on guest-host cells and color filters or polarizers.

4.1 Guest-Host Effects

Alternatively, a color LCD can be made by adding a pleochroic or dichroic dye molecule to the liquid crystal. A pleochroic is a molecule in which the absorption of light varies with the direction of the electric field vector, and hence is dependent upon the polarization. Different optical polarizations are absorbed in differing amounts. A dichroic is a pleochroic which produces plane polarized light through such selective absorption. A dichroic molecule is a long-chain molecule which absorbs radiation in a vibratory mode and transmits only the mutually perpendicular mode. For

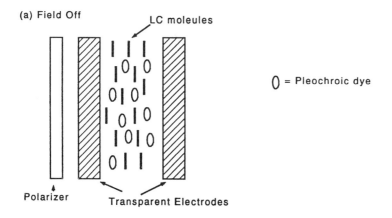

(a) Field Off   LC moleules

0 = Pleochroic dye

Polarizer   Transparent Electrodes

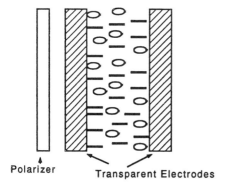

(b) Field On

Polarizer   Transparent Electrodes

Figure 11 Color liquid crystal operating with a pleochroic dye. In (a) both the dye and liquid crystal molecules are oriented perpendicular to the electric vector of the incident light, which is polarized by the front polarizer. The transmitted light has a spectral output which is determined by the absorption spectrum of the dye, and the cell appears colored. Under an electric field in (b), the molecules are tilted and the dye can no longer absorb. The cell then appears clear.

the device to work properly, the long-chain molecules of the dichroic must be aligned parallel to one another. Thus the dichroic molecules affect optical radiation in different ways, absorbing and transmitting depending on the polarization of the incident light. For a dichroic dye the absorption characteristics of the dye change with respect to the orientation of the incident polarization vector. If the long axis of the dye molecule is parallel to the electric vector of incident plane polarized light, absorption of the light takes place. However, the dye does not absorb all visible optical energies equally. Thus the spectral absorption for the dye has a characteristic shape, and the display emits the characteristic color of the dye. When the molecule is oriented with its axis perpendicular to the incident light, no absorption takes place and the light transmits unchanged.

If a dichroic dye is mixed with a liquid crystal in a supertwist or twisted nematic configuration, then the long chains of the dye will orient themselves with respect to the axis of the liquid crystal molecules. This "co-operative" ordering is illustrated in Fig. 11. With the electric field off, there is no tilt to either the liquid crystal or the dye molecules. A polarizer is attached to the entrance of the device and is oriented in such a way so the plane of polarization is perpendicular to the axis orientation of the molecules. The display then appears colored according to the transmittance of the dye. When an electric field is applied, the molecules tilt, the polarized light is transmitted unchanged, and the display appears white.

Besides use in twisted nematic cells, pleochroic dyes as guest-hosts have the potential to be used in liquid crystal displays requiring no polarizers. Double-layer liquid-crystal cells are "crossed"; that is, the dichroic dye molecules in each cell are oriented perpendicular to each other in the off state. Light is then blocked from passing through the two cells. In the on state the nematic liquid crystal molecules, plus the dichroic dye molecules, are oriented perpendicular to the electric field, uncrossing the polarization and allowing the light to pass. Such cells possess superior brightness and wider view angles as compared to twisted nematic cells.

5 MULTIPLEXING AND ADDRESSING

Liquid crystals can be formed in relatively large substrates for use in a variety of display applications, including portable television, laptop computers, and large-screen high-resolution projection television. A large transmission display can be prepared with a backlight for illumination and conductive electrodes on each substrate patterned to form the pixels. Figure 12 illustrates how pixels are formed from transparent electrodes patterned at right angles on the glass substrates. A pixel is defined by the area where the electrodes overlap. Nematic liquid crystal material is

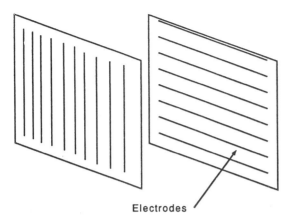

Electrodes

Figure 12 Example of how pixels are formed from transparent electrodes patterned on glass at right angles for a large multiplexed LCD display. Liquid crystal material is placed between the plates.

placed between the two substrates. A low or medium-resolution display can be prepared in this way, with conventional matrix addressing. One multiplexed signal drives each row and one drives each column. The difference between the row and column voltages equals the voltage experienced by the pixel. Usually, the rows are broken into halves in order to reduce the number of row access lines and to allow each pixel to remain on for longer times.

However, due to unintentional defects in the liquid crystal, selected pixels may crosstalk with unselected ones. For instance, when a column is addressed, charge leakage can turn on a pixel of the selected column but in an unselected row. Furthermore, if more rows are added, voltages must be reduced in order to maintain speed of the device, and the threshold voltages may not be precise nor uniform enough to make a reliable device.

As the technology progresses, new liquid crystal materials have been introduced with steeper electro-optic response curves. Lower voltage can then be used, and consequently more rows and columns have been added as the technology improved.

Other attempts to solve these problems have led to the development of *active matrix* addressing. In an active matrix system the drive voltage of the pixel is decoupled from the drive voltage of the row-column address system by placing a switch at the pixel that is activated by the row-column matrix. The switch is usually in the form of a thin-film transistor (tft). Thin-film transistors are fabricated in amorphous silicon, which is deposited in layers on the back surface of the LCD. Transistors are formed in the a-silicon, and the matrix address system is then patterned to select transistors rather than the LCDs themselves. The transistors then act as switches to activate the LCD pixels. Using this method, the pixel addressing is then independent of the value or steepness of the LCD threshold voltage, and is instead dependent upon the threshold voltage of the transistor switch, which is more uniform across the display. Alternatively, instead of transistor switches, active matrix LCDs have been made using addressed glass columns filled with plasma gas fixed below threshold. The plasma-assisted display then acts to access the liquid crystal pixel.

Figure 13 illustrates an aluminum-gate tft on a glass substrate. The source of the transistor is connected to the LCD pixel, power is applied to the drain, and the gate is accessed via a matrix of thin aluminum films. The transistor's metal-insulator gate is Al-Al$_2$O$_3$. The aluminum gate is anodized to form the Al$_2$O$_3$. The use of low-resistivity aluminum helps reduce the delay of the gate busline.

On top of the anodized Al$_2$O$_3$, are layers of SiN and amorphous Si:H films deposited by plasma-enhanced chemical vapor deposition. The SiN layer forms a barrier interface between the aluminum oxide and the amorphous Si. The drain, source, and channel are patterned in the amorphous silicon [4].

A less expensive alternative to the active matrix is the *active addressed* display. The active-addressed system uses a regular passive liquid crystal matrix, but uses a complex row-column addressing system in order to avoid crosstalk. Complex electronics must be used to provide addressing, but the result is an overall high-speed, relatively inexpensive display.

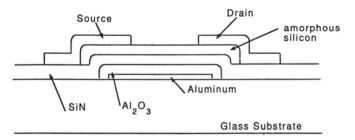

Figure 13 Cross-sectional view of a silicon thin-film transistor prepared on a glass substrate with an Al$_2$O$_3$/SiN double layer used as a gate insulator. (From Ref. 4.)

6 OPTICALLY ADDRESSED LCDS

The potential uses of liquid crystals in adaptive optics have long been recognized [5]. Simple focusing structures can be made by using small rectangular electrodes on liquid crystal cells that mimic Fresnel cylindrical lenses. Spatial light modulators (SLMs) are used in optical processing, optical correlation, and optical computing. Spatial phase modulators [6,7] and intensity modulators can be constructed from liquid crystals; the latter is more commonly referred to as a "light valve." The liquid crystal light valve has applications in large-screen projection. The first successful commercial product was developed by Hughes Aircraft in the mid-1970s.

If a photoconductor is added to a liquid crystal display, an optically addressed device can be made. An input light signal can be used to control an output light signal. Since both photoconductor and liquid crystal can be prepared as two-dimensional arrays, entire images can be used to control output images. Applications of such devices include image amplifiers, image storage devices, and optically addressed spatial light modulators. The latter is used to convert an image of incoherent light into one of coherent light.

In such optically addressed devices, the liquid crystal portion operates on one of the principles described in the chapter: The device controls a secondary source of light through optical polarization rotation or reflection. Electrical control of the device, however, comes not from an externally supplied electrical stimulus, but from an electrical stimulus from the photoconductor.

An example is shown in Fig. 14. A write beam is incident on the underside of the device in the figure and strikes a photoconductor layer which is separated by electrodes, forming a capacitor with the photoconductor as the dielectric. The capacitor formed is externally precharged. Bright areas of the write image are absorbed by the photoconductor, and the capacitor discharges in that region of the device. An electrical signal dependent upon the dark areas of the image is then directed onto the liquid crystal matrix. A light-blocking layer restricts the write beam from continuing past the photoconductor.

On the top side of the device in the figure, the read beam is reflected through the liquid crystal and off of a dielectric mirror sandwiched in the device. The incident read beam is linearly polarized, and the liquid crystal acts to rotate the plane of polarization of the reflected read beam through an output polarizer. The electrical charge from a dark area of the write beam activates the liquid crystal under that area, which prevents rotation of the read beam polarization plane, and the light is blocked by the exit

Incident read beam   Reflected read beam

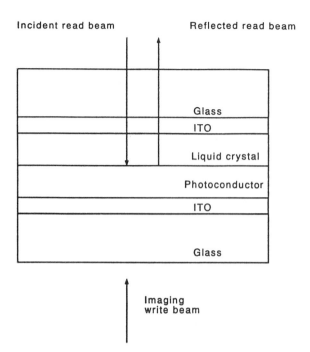

Glass

ITO

Liquid crystal

Photoconductor

ITO

Glass

Imaging
write beam

Figure 14 Example of an optically addressed spatial light modulator formed from a liquid crystal and photoconductor sandwiched between two electrodes. A write beam incident on the photoconductor provides the stimuli to activate the liquid crystal. The latter then acts to modulate a read beam.

polarizer. As shown in the figure, the device then produces a positive of the image.

The input image to the light valve is often generated by a cathode ray tube [8]. A lens can be used to image the CRT onto the light valve. Alternatively, a fiber-optic faceplate can be attached to the cathode ray tube, and fiber optics can be directed to the liquid crystal light valve. This allows direct coupling of light from the CRT to the light valve. Such a combination works well for large-screen applications, including high-definition applications containing more than 1000 scan lines.

The LCLV can also be addressed with a laser [9]. If a laser is used to write to the valve in Fig. 14, a modulated laser scans the light valve and forms the image. Another type of laser-addressed light valve makes use of the differences in thermodynamics and kinetics of liquid crystal phases.

This type of liquid crystal is shown in Fig. 15. For this device the photo-conductor is replaced with an absorptive layer that absorbs optical radiation from the laser. Heat generated in the absorptive layer is transferred to the liquid crystal layer. As the liquid crystal heats, it changes from the smectic phase to the isotropic phase. If a voltage is applied to the pixel when the liquid crystal cools, it orients according to the electric field, and the pixel appears transparent. If, however, no voltage is applied to the pixel, the liquid crystal is unoriented as it returns to the smectic phase, and the pixel is light scattering and appears dark. Laser-assisted addressing is very slow, but the image is contained for many hours, and such a display is capable of high resolution.

If the write operation is separated in time from the read operation, then the device of Fig. 14 or 15 may function as an optical storage device. If the read light is of higher intensity than the incident write light, the device serves as an image amplifier. Finally, if the write image is incoherent and the read light is from a coherent source, the device serves as an incoherent-to-coherent image converter.

Similar spatial light modulators can be prepared by using reversed-biased *p-i-n* diode sandwich layers, or charge-coupled devices in place of the photoconductor, to convert the write beam into a spatially variant electrical signal. This electrical signal can then be used to produce the image on the liquid crystal matrix.

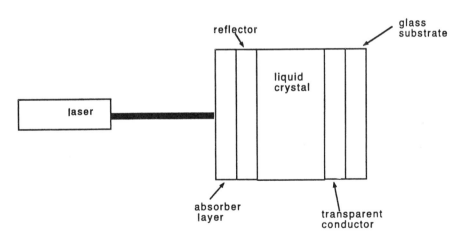

Figure 15   Laser-addressed liquid crystal light valve which makes use of an absorptive layer to produce local temperature variations in the liquid crystal layer.

7 POLYMER LCDS

Liquid crystals can also be prepared as polymers or macromolecules. These are long-chain high-molecular-weight molecules made from small-molecular-weight liquid crystals. An example is shown in Fig. 16. The rodlike, small-molecular-weight liquid crystals are added to a "backbone" to form a liquid crystal macromolecule. The backbone is usually a carbon chain, and such a structure is termed a side-chain liquid crystal polymer.

Liquid crystal macromolecules exhibit similar electrical characteristics to their smaller-weight side chains. They orient themselves under an electric field and act as polarization rotators to incident light. They form nematic, smectic, and cholesteric phases, only at higher temperatures than their lower-molecular-weight counterparts. Because of the increased molecular size, they tend to react more slowly than regular LCDs. In addition, they have certain fabrication advantages as do commercial non-liquid-crystal polymers.

Polymer LCDs also form another crystalline phase not shared by low-molecular-weight LCs. As cooled, a regular liquid crystal will go from isotropic liquid to liquid crystal to crystalline solid. A polymer liquid crystal will go from liquid to liquid crystal to a glassy state. The glassy state is characteristic of polymers. Upon transition to the glassy state, however, the polymer molecules are frozen into the same relative positions they enjoyed in the liquid crystal state. It is then possible to make an isotropic glass which has some ordering to its macromolecules. The device does

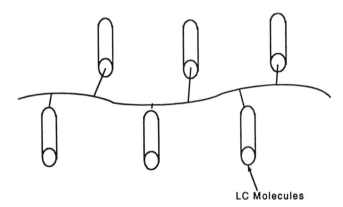

LC Molecules

Figure 16 Structure of a liquid crystal polymer. The small-molecular-weight liquid crystals, which are rod-shaped molecules as depicted in the illustration, are attached as side chains to a carbon backbone to form a liquid crystal polymer.

have a memory, where a previous state oriented under the appropriate electric field can be captured indefinitely. The glassy liquid crystal polymers have superior mechanical strength to low-molecular-weight materials and can be fabricated in sheet form.

7.1 Polymer-Dispersed Liquid Crystals

A new type of liquid crystal display can be made by forming micron-sized droplets of liquid crystal in a transparent polymer matrix [10, 11]. The droplets are formed by a phase separation process from a mixture of prepolymer and liquid crystal. The phase separation can be forced when the prepolymer polymerizes or by solvent evaporation or thermal quenching. The result of the phase separation process are near-spherical liquid crystals dispersed within the polymer.

A schematic of such a polymer droplet is shown in Fig. 17. With no external electric field present, the molecular orientations of the droplets are determined by any surface inhomogeneities within the polymer inclusions. When an electric field is applied, it reorients the liquid crystal within these inclusions and thus changes the refractive index difference between

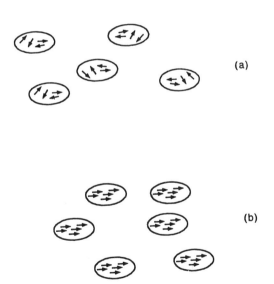

Figure 17 Example of liquid crystal polymer droplets. With no electric field (a) the orientations of the liquid crystal domains are determined by the surface inhomogeneity of the droplet. Under an electric field (b) the liquid crystal within the droplet orients in the direction of the applied field.

the liquid crystal and the polymer matrix. The dispersed liquid crystal then makes use of a scattering mode of operation and can be switched from a cloudy light-scattering state to a transparent state.

Since no polarizers are needed for dispersed liquid crystal displays, more light can be transmitted through a cell, and the cell has applications in high-intensity projection devices. Other advantages of using dispersed liquid crystals include the relative ease of processing polymer films, which can be fabricated in giant sheets, as compared to fabrication of twisted nematic cells.

REFERENCES

1. P. G. de Gennes, *The Physics of Liquid Crystals*, Clarendon Press, Oxford, 1974, Chapter 3.
2. HI. de Vries, Rotatory power and other optical properties of certain liquid crystals, *Acta Crystallogr.*, *4*, 219 (1951).
3. C. H. Gooch and H. A. Tarry, The optical properties of twisted nematic liquid crystal structures with twist angles $\leq 90°$, *J. Phys. D: Appl. Phys.*, *8*, 1575 (1975).
4. H. Yamamoto, H. Matsumara, K. Shirahashi, M. Nakatani, A. Sasano, N. Konishi, K. Tsusui, and T. Tsukada, A new a-Si TFT with Al_2O_3/SiN double-layered gate insulator for 10.4 inch diagonal multicolored display, *Tech Digest Int. Electron Dev. Meeting* 851, 1990.
5. A. Purvis, G. Williams, N. J. Powell, M. G. Clark, and M. C. K. Wiltshire, Liquid crystal phase modulators for active micro-optic devices, in *Liquid-Crystal Devices and Materials*, *Proceedings SPIE*, *1455* (P. S. Drzaic, and U. Efron, eds.), SPIE, 1991, p. 145.
6. J. L. Horner and P. D. Gianino, Phase-only matched filtering, *Appl. Opt.*, *23*, 812 (1984).
7. J. D. Armitage and D. K. Kinell, Miniature spatial light modulators, in *Advances in Optical Information Processing IV*, *Proceedings of the SPIE*, *1296*, p. 158 (1990).
8. W. P. Bleha, Progress in liquid crystal light valves, *Laser Focus/Electro-Optics*, *Oct.* (1983).
9. A. G. Dewey, Laser-addressed liquid crystal displays, *Opt. Eng.*, *23*, 230 (1984).
10. J. W. Doane, N. A. Vaz, B. G. Wu, and S. Zumer, Field controlled light scattering from nematic microdroplets, *Appl. Phys. Lett.*, *48*, 269 (1986).
11. J. W. Doane, *Liquid Crystals*: *Applications and Uses* (B. Bahadur, ed.), World Scientific, 1990, Chapter 14.

BIBLIOGRAPHY

F. D. Saeva, ed. *Liquid Crystals*: *The Fourth State of Matter*, Marcel Dekker, 1979.

F. J. Kahn, The molecular physics of liquid crystal devices, *Phys. Today 35*, 66 (1982).

S. D. Jacobs, ed. *Selected Papers on Liquid Crystals for Optics*, SPIE Optical Engineering Press, 1992.

M. Schadt, The history of the liquid crystal display and LC-material technology, *Liquid Crystals*, *5*, 57 (1989).

Index